FROM
WHITE DWARFS
TO BLACK HOLES

FROM
WHITE DWARFS
TO BLACK HOLES

The Legacy of S. Chandrasekhar

Edited by

G. SRINIVASAN

THE UNIVERSITY OF CHICAGO PRESS

CHICAGO AND LONDON

G. Srinivasan is a member of the Raman Research Institute. He is also the editor of the *Journal of Astrophysics and Astronomy*, published by the Indian Academy of Sciences.

The University of Chicago Press, Chicago 60637
The University of Chicago Press, Ltd., London
© 1996, 1999 by The Indian Academy of Sciences
All rights reserved. Published 1999
08 07 06 05 04 03 02 01 00 99   5 4 3 2 1

ISBN: (cloth) : 0-226-76996-8

The essays in this book were originally published by the Indian Academy of Sciences in 1996 in a special issue of the *Journal of Astrophysics and Astronomy*.

Library of Congress Cataloging-in-Publication Data

From white dwarfs to black holes : the legacy of S. Chandrasekhar /
  [edited by] G. Srinivasan.
      p.   cm.
   Includes bibliographical references.
   ISBN 0-226-76996-8 (alk.paper)
   1. Astronomy. 2. Chandrasekhar, S. (Subrahmanyan), 1910–
3. Astrophysicists—United States—Biography. I. Chandrasekhar, S.
(Subrahmanyan), 1910–   . II. Srinivasan, G. (Ganesan), 1942–   .
QB51.F734   1999
523.01'092—dc21
                                                                98-48661
                                                                CIP

∞ The paper used in this publication meets the minimum requirements of the American National Standard for Information Sciences—Permanence of Paper for Printed Library Materials, ANSI Z39.48-1992.

Contents

Preface

Subrahmanyan Chandrasekhar was a legend in his own times. He was one of the colossal figures of 20th century science. Very few can match his sustained productivity at the highest level for 65 years. In his achievements and in his intense scholarship he has been compared with Lord Rayleigh and Poincaré.

Chandra, as he was known the world over, was born on the 19th of October 1910. He was born into a cultured and very gifted family. He made his appearance on the international scientific scene at the young age of 18 when he was a B.Sc. student at the Presidency College in Madras. The year was 1928. The Compton Effect was the hot topic in physics. During that summer Chandra visited his uncle C. V. Raman in Calcutta. There the Indian Association for the Cultivation of Science was buzzing with excitement. Just a couple of months earlier Raman had discovered the famous effect for which he was to receive the Nobel Prize two years later. It was a most stimulating experience for young Chandra to spend the summer months in the company of the distinguished group working with Raman. Soon after this visit he wrote his first scientific paper, entitled "Thermodynamics of Compton scattering with reference to the interiors of stars." Later that year the great German physicist Arnold Sommerfeld visited the Presidency College. It is from him that Chandra learned of the discovery of the new statistics by Fermi and Dirac, and the application of this new statistics to the electron gas in metals by Sommerfeld himself. Inspired by this, Chandra looked for another problem to which to apply the new statistics. The newly discovered Compton Effect suggested an interesting problem. Within a couple of months his researches led him to write his second paper, entitled "Compton scattering and the new statistics." He sent this paper to R. H. Fowler in Cambridge, requesting him to communicate it to an appropriate journal. Fowler got young Nevill Mott to referee it and communicated it to the *Proceedings of the Royal Society of London.*

The following year Chandra became aware of the seminal paper by Fowler invoking the Fermi-Dirac statistics to explain the stability of white dwarfs. Straightaway he built on Fowler's idea and derived the celebrated mass-radius relation for white dwarfs. In 1930 he went to Cambridge to pursue his research under the guidance of Fowler. During the next five years—perhaps the most brilliant phase of his long and illustrious career—he obtained several fundamental results which are now recognized to be at the base of the present revolution in high energy astrophysics.

But unfortunately they were not seen to be so at that time. The main reason was that Sir Arthur Eddington in Cambridge, the most distinguished astrophysicist at that time, did not believe in the startling results obtained by Chandra. Indeed this led to a major controversy which was to have a lasting effect on Chandra. Faced with the enormous pressures of finding himself at the center of a major controversy with the leading astrophysicist in the world, Chandra decided to leave the subject of stellar structure altogether and move on to other things. He also decided to leave Cambridge. In 1937 he joined the Yerkes Observatory of the University of Chicago. Otto Struve, the Director of the Observatory, was in the process of hiring some of the world's most brilliant young astronomers and astrophysicists, and Chandra was one of them. He was associated with the University of Chicago till his death.

During the intervening six decades his research interests spanned an incredibly wide spectrum. But there was a distinct pattern. There were eight distinct periods in his life. They are, briefly:

1. Stellar structure, including the theory of white dwarfs [1929–39].

2. Stellar dynamics, including the theory of Brownian motion [1939–43].

3. Theory of radiative transfer, the theory of illumination and the polarization of sunlit sky, the theories of planetary and stellar atmospheres, and the quantum theory of the negative ion of hydrogen [1943–50].

4. Hydrodynamic and hydromagnetic stability [1952–61].

5. The equilibrium and the stability of ellipsoidal figures of equilibrium [1961–68].

6. The general theory of relativity and relativistic astrophysics [1962–71].

7. The mathematical theory of black holes [1971–83].

8. The theory of colliding gravitational waves and non-radial perturbations of relativistic stars [1983–95].

Soon after leaving Cambridge Chandra gathered together all his results on stellar structure and published it as a book entitled *An Introduction to the Study of Stellar Structure*. This book is universally acclaimed as a masterpiece of the first rank. Next he turned to the problem of the dynamics of star clusters. During the period 1939–43 he contributed mightily to this subject. Unlike other astronomers, who were worrying about specific problems, he approached stellar dynamics as a discipline in itself, bringing forth and trying to solve its own theoretical problems. Thus he was able to formulate *"certain abstract problems which appear to have an interest for general dynamical theory even apart from the practical context in which they arise."* His classic monograph *Principles of Stellar Dynamics* was published in 1942.

The next period, 1943–50, was devoted to an investigation of the extremely difficult problem of radiative transfer in stellar and planetary atmospheres. Although this problem originated in Lord Rayleigh's investigations in 1871, the formulation of the fundamental equations and their solution had to wait till Chandra turned his attention to it. Approaching the problem of radiative transfer as a branch of mathematical physics, he developed novel techniques to generalize and exploit certain principles of invariance which had been formulated by Victor Ambartsumian. During this period he also made many important contributions to the quantum theory of the negative ion of hydrogen. In 1950 he published his monumental book *Radiative Transfer*. Chandra often said that this phase of his career gave him the greatest satisfaction.

During the next decade he focused his attention almost entirely on the statistical theory of turbulence and of hydrodynamic and hydromagnetic stability. He realized that unless substantial progress was made in these extremely difficult branches of physics, many exciting astrophysical problems could not be addressed. The year 1961 saw the publication of his mammoth book *Hydrodynamic and Hydromagnetic Stability*.

In the beginning of the 1960s he was asked to give a series of four lectures at Yale University, and he chose for the topic of his lectures "Rotation of astronomical bodies." While preparing for these lectures he became aware of the classic works of giants like Maclaurin, Riemann, Jacobi, and others. He devoted the next 5 or 6 years to cleaning up this extremely difficult field left incomplete by some of the greatest figures in the history of mathematics. In the epilogue of the book he wrote:

> But the subject, nevertheless, had been left in an incomplete state with many gaps and omissions and some plain errors and misconceptions. It seemed a pity that it should be allowed to remain in that destitute state. Whether the effort expended in its rehabilitation was worth the time, the author cannot presume to judge.

These investigations led to the publication in 1969 of his book *Ellipsoidal Figures of Equilibrium*.

Around 1962 he got interested in general relativity. The great revolution in relativistic astrophysics was just below the horizon. Chandra was apprehensive about entering this field dominated by young stars. But the first problem he chose to attack suited his taste, temperament, and talent.

He went back to a problem he had initiated in the 1930s while still in Cambridge: the stability of rotating and vibrating stars within the framework of general relativity. He had started this work in collaboration with von Neumann when the latter visited Cambridge. He returned to this problem and considered the following two questions: (1) What is the influence of general relativity on the critical adiabatic index for the stability against radial perturbations of a non-rotating star? and (2) Do dissipative effects of gravitational radiation reaction induce secular instabilities in rotating stars? Both these questions were well posed and were amenable to the kind of analysis in which he was the supreme master. This led to a series of extremely important papers which acquired great significance with the discovery of pulsars, quasars, and active galactic nuclei.

Next he turned towards the theory of gravitational radiation. The problem he set out to solve was the following: If one developed Einstein's theory of gravitation as a series expansion in powers of v/c, with Newton's theory as the zero order theory, then in what order of v/c will the existence of gravitational radiation become important? In a remarkably short span of less than two years Chandra and one of his students demonstrated for the first time the existence of gravitational radiation in the 2 1/2 post-Newtonian approximation. Curiously, for the first time in his career, he did not pause to write a comprehensive book before moving on to an entirely different line of investigation.

The reason was quite simple. Just around that time relativistic astrophysics was coming of age. Penrose and Hawking had published their fundamental papers on singularity theorems. Pulsars had been discovered. Convincing arguments had been made that quasars and active galactic nuclei must harbor massive black holes. This led to a revival of activity in general relativity. Some of the most brilliant students chose to work in this area. It was therefore natural that Chandra, too, should enter this field, for, after all, the fundamental discoveries he had made in the 1930s had led one inescapably to the concept of black holes and singularities. The problem he chose to concentrate on concerned the stability of black holes against external perturbations such as gravitational and electromagnetic waves. He worked incredibly hard during this period, not withstanding serious health problems, and wrote a series of very long and technically difficult papers. Finally he wrote a monumental book: *The Mathematical Theory of Black Holes*. This book, published in 1983, became an instant classic.

Chandra was 73 years old at this stage. Many of his colleagues wondered whether he would choose to retire, but he continued with unabated enthusiasm. During the last ten years of his life he concentrated on two difficult problems: the collision of gravitational waves, and non-radial perturbations of relativistic stars. He began his research in the 1930s with the study of the stability of relativistic stars. Sixty-five years later he ended his incredibly productive career on the same note.

The final project he undertook was to write a commentary on Newton's *Principia*. Like many, Chandra regarded this book as the greatest intellectual achievement of the human mind. Like everything else he had started, he completed this project successfully, and his book *Newton's Principia for the Common Reader* was published just a few weeks before he died.

Chandra's unique attitude to science must be clear from this pattern. The most distinctive character of his research was that after having carefully chosen an important problem which was amenable to his talent and temperament, and after contributing mightily to it, he wrote a monumental treatise in which the subject was presented from a unified perspective which was his own. About this attitude of striving to understand things in his own way, within his own framework, he said:

> After the early preparatory years my work has followed a certain pattern motivated, principally, by quest after perspectives. In practice, this quest has consisted of my choosing (after trials and tribulations) a certain area which appears amenable to cultivation and compatible with my taste, abilities, and temperament. And when after some years of study I feel that I have accumulated sufficient body of knowledge and achieved a view of my own, I have the urge to present my point of view *ab initio* in a coherent account with order, form and structure.

Along with research, teaching was an integral part of Chandra's life. He prepared for his lectures with painstaking thoroughness, and they were delivered in a masterly fashion; every step and every argument would be written on the board in his beautiful handwriting. More than fifty students worked with him for their Ph.D. degrees. He considered his collaboration with young scientists an essential part of his scientific style, and much more important than his collaboration with giants like Enrico Fermi and John von Neumann. He also found it very inspiring to work with young people. This was particularly so in the last thirty years. He once said:

> I consider myself very fortunate in having made up my mind to do relativity. Among other things, for the first time, certainly after the early forties, I felt I was working in an area in which others were far more equipped than I was. I thought I had a chance of having a close scientific proximity with people of the highest caliber. ... I feel once more rejuvenated, once again with young people tremendously bright, tremendously exciting.

He had experienced this once before—when he spent a year in Copenhagen in the company of extraordinarily brilliant young people working with Neils Bohr. He was 22 years old then. Very few scientists experience that excitement twice in their lives. When Chandra decided to work in general relativity he was nearly 60 years old; he did so because this would give him an opportunity to experience once more the excitement of working with brilliant young people.

One of Chandrasekhar's outstanding contributions to the astronomical community was to build up the *Astrophysical Journal*. He was its managing editor from 1952 to 1971. When he assumed the editorship the *Journal* published six issues a year totaling approximately 1,000 pages. In 1970 it became twenty-four issues a year totaling over 12,000 pages. In 1967 he started the *Astrophysical Journal Letters* section to be able to publish short accounts of important discoveries faster. He was extremely demanding as an editor, and his style of functioning was at times resented. But no one can deny that his uncompromising attitude was solely responsible for the reputation enjoyed at present by the *Astrophysical Journal* as the leading journal in the field.

An important aspect of Chandra's science, particularly in his later years, was his quest for beauty in science. His concept of beauty in science was based on the following two criteria:

> There is no excellent beauty that hath not some strangeness in the proportion!

> —Francis Bacon

> Beauty is the proper conformity of the parts to one another and to the whole.

> —Werner Heisenberg

Chandra stressed that the experience of beauty in science is not limited to the context of great ideas by great minds.

> This is no more true than the joys of creativity are restricted to a fortunate few. They are, indeed, accessible to each one of us provided we are attuned to the perspective of strangeness in the proportion and the conformity of the parts to one another and to the whole. And there is satisfaction also to be gained from harmoniously organizing the domain of science with order, pattern and coherence....

Not surprisingly, Chandra received numerous awards. These include the Bruce Medal (Astronomical Society of the Pacific), the Royal Medal (Royal Society of London), the Srinivasa Ramanujan Medal (Indian National Science Academy), the National Medal of Science (United States of America), the Copely Medal (Royal Society of London), and the Nobel

Prize for Physics (Royal Swedish Academy). The Indian Academy of Sciences (founded in 1934) is naturally proud of the fact that Chandrasekhar was one of its Founding Fellows.

Chandra admired the lengthy obituaries that appeared, mostly in the nineteenth century, by eminent scientists about eminent scientists. Following this style, he himself carefully analyzed and assessed the contributions of Fowler, Milne, Eddington and Struve. The range of Chandra's contributions is so vast that no one person in the physics or astronomy community can undertake the task of commenting on his achievements. In this memorial volume, a number of eminent scientists have collectively undertaken this daunting task.

Chandra himself was quite modest in assessing his own contributions. This was but natural since his standards were very high. In a memorable lecture delivered in 1985 during the Golden Jubilee Meeting of the Indian Academy of Sciences he said:

> The pursuit of science has often been compared to the scaling of mountains, high and not so high. But who amongst us can hope, even in imagination, to scale the Everest and reach its summit when the sky is blue and the air is still, and in the stillness of the air survey the entire Himalayan range in the dazzling white of the snow stretching to infinity? None of us can hope for a comparable vision of nature and of the universe around us. But there is nothing mean or lowly in standing in the valley below and awaiting the sun to rise over Kinchinjunga.

He did not seek or greatly value praise from his contemporaries. But he wished they would understand and appreciate his motivation for the pursuit of science. As regards his achievements he often quoted from a letter he had received from his friend Edward Arthur Milne in the 1930s:

> On an occasion, now more than 50 years ago, Milne reminded me that posterity, in time, will give us all our true measure and assign to each of us our due and humble place; and in the end it is the judgement of posterity that really matters. And Milne further added: He really succeeds who perseveres according to his lights, unaffected by fortune, good or bad. And it is well to remember that there is in general no correlation between the judgement of posterity and the judgement of contemporaries.

It may well be. But there is no doubt that posterity will regard Subrahmanyan Chandrasekhar, along with Karl Schwarzschild and Arthur Stanley Eddington, as the most distinguished astrophysicists of the twentieth century.

G. Srinivasan

1

Stars: Their Structure and Evolution

G. Srinivasan

1.1 Introduction

The subject of astrophysics began with the study of the stars. It may be recalled that the positivist philosophers who were so influential in European
thinking had asserted that it was in the nature of things that one can never
know what the stars are. And yet, with Fraunhofer's discovery of dark lines
in the spectrum of the Sun and the stars, and their subsequent explanation in terms of atomic absorption lines, a major scientific revolution had
occurred—a question that appeared meaningless within the framework of
science had acquired a meaning. Lane, Kelvin, and Helmholtz laid the
foundations for the theory of stars towards the end of the 19th century.
But the credit for constructing a remarkably successful theory of the stability and equilibrium of stars must go to Sir Arthur Eddington. His book
The Internal Constitution of the Stars, published in 1926, is undoubtedly
one of the greatest masterpieces of the 20th century.

It is from this book that young Chandrasekhar learned about the theory
of the stars. The year was 1928, and he was an undergraduate student in
the Presidency College in Madras. The newly discovered Compton Effect
was much in the news, and was the subject of his first scientific publication, entitled "Thermodynamics of the Compton Effect with reference to
the interior of the stars" (Chandrasekhar 1928). He was 18 years old then.
That same year he learned about the discovery of the Fermi-Dirac statistics
from Arnold Sommerfeld, who happened to visit Madras. Straightaway he
applied the new statistics to Compton scattering, and this paper was communicated to the Royal Society by R. H. Fowler (Chandrasekhar 1929).
During the next ten years he made monumental contributions to the theory of stellar structure and stellar evolution. Much of it is summarized in
his classic book entitled *An Introduction to the Study of Stellar Structure*,
published in 1939. Almost immediately after writing this book, he decided
to leave the field, and turned his attention to problems in Stellar Dynamics. This decision was primarily due to the fact that his epoch-making

discoveries, instead of being lauded, stirred up a great controversy, largely on account of Eddington rejecting them. In this article we shall attempt to recall some of his pioneering contributions to the theory of the stars. This is undoubtedly a daunting task. Fortunately, Chandrasekhar himself had made a selection of his most important papers from this period for inclusion in volume 1 of his *Selected Papers*. We shall choose a few of them, explain their content, and trace their impact on subsequent developments.

1.2   The theory of white dwarfs

Chandrasekhar's first significant papers were devoted to the theory of white dwarfs. Following Chandrasekhar's own style of writing, we shall digress for a moment to enable one to place his contributions in proper perspective.

The discovery of white dwarfs, such as the companion of Sirius, with mean densities of the order of 10^5—10^6 g cm^{-3}, appeared to spell trouble for the enormously successful theory due to Eddington. As Eddington himself put it in his book:

> I do not see how a star which has once got into this compressed state is ever going to get out of it. ... It would seem that the star will be in an awkward predicament when its supply of subatomic energy fails.

The difficulty may be explained as follows. The electrostatic energy E_v per unit volume of an assembly of completely ionized atoms (with nuclear charge Z) is given by

$$E_v = 1.32 \times 10^{11} Z^2 \rho^{4/3}, \tag{1.1}$$

where ρ is the mass density. The kinetic energy E_{kin} per unit volume of the free particles (under the assumption that it is a *perfect gas*) is given by

$$\begin{aligned} E_{\text{kin}} &= \frac{3}{2}\frac{k}{\mu m_H}\rho T \\ &= \frac{1.24 \times 10^8}{\mu}\rho T, \end{aligned} \tag{1.2}$$

where μ is the mean molecular weight. If the external pressure (in this case gravitational pressure) were removed, the matter can expand and return to its original state of normal atoms only if

$$E_{\text{kin}} > E_v, \tag{1.3}$$

or if

$$\rho < \left(0.94 \times 10^{-3}\frac{T}{\mu Z^2}\right)^3. \tag{1.4}$$

The point underlying Eddington's remarks is that this inequality is violated under the conditions that obtain in white dwarfs. This paradox was resolved in 1926 by Fowler in one of the most prescient papers in the astronomical literature. Fowler argued that at high densities electrons will be highly degenerate and therefore their kinetic energy and pressure should be calculated not according to Boyle's law, but according to the newly discovered quantum statistics. According to the Fermi-Dirac statistics the pressure and kinetic energy of a highly degenerate electron gas are given by

$$P = \frac{1}{20} \left(\frac{3}{\pi}\right)^{2/3} \frac{h^2}{m} n^{5/3}, \tag{1.5}$$

and

$$
\begin{aligned}
E_{\text{kin}} &= \frac{3}{40} \left(\frac{3}{\pi}\right)^{2/3} \frac{h^2}{m} n^{5/3} \\
&= 1.39 \times 10^{13} \left(\rho/\mu\right)^{5/3}, \tag{1.6}
\end{aligned}
$$

where n and ρ are the number density and mass density, and μ is the mean-molecular weight. Fowler's point was that the inequality (1.3) will easily be satisfied if one uses the quantum statistical expression for the kinetic energy instead of its classical value.

Chandrasekhar became aware of Fowler's seminal paper in 1929, and immediately applied the theory of polytropic gas spheres to the new equation of state and derived the mass-radius relation for completely degenerate white dwarf configurations. This historic paper was communicated to the *Philosophical Magazine* by Fowler. To recall briefly, the Fermi-Dirac pressure of a degenerate electron gas (at absolute zero of temperature) is given by a polytropic equation of state

$$P = K\rho^{5/3}. \tag{1.7}$$

This corresponds to a polytropic index $n = 3/2$ (where $5/3 = 1 + 1/n$). Using the theory of equilibrium states of polytropic gas spheres, Chandrasekhar obtained the following relations between the masses and radii of white dwarfs, as well as their mean density:

$$\frac{M}{M_\odot} = \frac{2.14 \times 10^{28}}{\mu^5} \cdot \frac{1}{R^3} \tag{1.8}$$

$$\rho = 2.162 \times 10^6 \left(\frac{M}{M_\odot}\right)^2. \tag{1.9}$$

To quote from this early paper (Chandrasekhar 1931a):

(i) the radius of a white dwarf is inversely proportional to the cube root of the mass;

(ii) the density is proportional to the square of the mass;

(iii) the central density would be six times the mean density.

Thus, according to this theory, *all* stars, regardless of their mass, will end their lives peacefully as white dwarfs.

1.2.1   The limiting mass

Soon after completing this work Chandrasekhar set out to Cambridge to continue his research under the guidance of R. H. Fowler. During the voyage he began to worry about the effects of special relativity on the conclusions he had just arrived at. He concluded that at electron densities $> 6 \times 10^{29}$ cm^{-3} one must use a modified form for the pressure of an electron gas. In the extreme relativistic case (when the rest mass of the electrons can be neglected) the pressure is given by

$$
\begin{aligned}
P &= \frac{1}{8} \left(\frac{3}{\pi} \right)^{1/3} \cdot hc \cdot n^{4/3} \\
&= K_2 \rho^{4/3}.
\end{aligned}
\tag{1.10}
$$

Once again applying the theory of polytropic gas spheres (this time for an equation of state with a polytropic index $n = 3$) he derived the relation

$$
\left(\frac{GM}{M'} \right)^2 = \frac{(4K_2)^3}{4\pi G},
\tag{1.11}
$$

where $M' = 2.7176$. With an assumed value of 2.5 for the mean molecular weight (the "canonical value" in 1930) this yielded

$$
M = 1.822 \times 10^{33} \text{ g} \approx 0.91 M_\odot.
\tag{1.12}
$$

More generally

$$
M = 0.197 \left[\left(\frac{hc}{G} \right)^{3/2} \cdot \frac{1}{m_H^2} \right] \frac{1}{\mu_e^2} = 5.76 \mu_e^{-2} M_\odot,
\tag{1.13}
$$

where μ_e is the mean molecular weight per electron. Thus a fully degenerate star, in the extreme relativistic limit, has a *unique mass!* Chandrasekhar concluded that this must represent the *maximum mass of an ideal white dwarf* (Chandrasekhar 1931b).

At this stage Chandrasekhar was not aware of the work of W. Anderson (1929) and E. C. Stoner (1929, 1930) who had independently investigated this problem. Fowler drew his attention to these papers upon his arrival in Cambridge. Stoner's approach was more heuristic. In his 1929 paper

Stoner had derived the *limiting* density for white dwarfs, using the following argument. The number of electrons with momenta within a definite range cannot exceed a certain maximum. Any increase in the density involves an increase of energy. In the limiting case, at absolute zero, the star can contract until the decrease in gravitational energy becomes insufficient to balance the increase of kinetic energy of electrons. The limiting density corresponds to the value of n when

$$\frac{d}{dn}\left(E_{\text{kin}} + E_{\text{grav}}\right) = 0. \tag{1.14}$$

This yielded a value of density approximately twice the *average* density calculated by Chandrasekhar using the more rigorous approach of hydrostatic equilibrium.

Historically, Anderson (1929) was the first to appreciate that at high densities special relativistic effects would have to be taken into account. In particular, he isolated the fundamental result that as the density increases the mass will approach a "limiting mass." Stoner (1930) improved upon this result (but confining himself to his earlier framework of considering the energetics of homogeneous spheres) and derived a limiting mass of 2.19×10^{33} g. Curiously, neither Anderson nor Stoner pursued this further!

The credit for elucidating the significance of the limiting mass must go solely to Chandrasekhar. Faced with the skepticism of R. H. Fowler and E. A. Milne, he puzzled over this intriguing result. But by October of that year (1930) it became clear to him that what was happening was that the relation $R \propto M^{-1/3}$ given by the nonrelativistic theory was modified by the inclusion of relativistic effects in the following way. Consider an approximation in which a white dwarf consists of a nonrelativistically degenerate "envelope" (in which the pressure $\propto \rho^{5/3}$) and a "core" (in which the pressure $\propto \rho^{4/3}$). In this approximation Chandrasekhar showed (1931c) that "the completely relativistic model, considered as a limit of this composite series is a point mass with $\rho_c = \infty$!"

Armed with this insight, Chandrasekhar proceeded to work out a complete theory which allowed for the effects of special relativity in an exact manner. For this he wrote the equation of state in a parametric form

$$P = Af(x), \quad \rho = n\mu_e m_H = Bx^3, \tag{1.15}$$

where

$$A = \frac{\pi m^4 c^5}{3h^3}, \quad B = \frac{8\pi m^3 c^3 \mu_e m_H}{3h^3}, \tag{1.16}$$

and

$$f(x) = x\left(x^2 + 1\right)^{1/2}\left(2x^2 - 3\right) + 3\sinh^{-1} x. \tag{1.17}$$

The above expression for the pressure approximates the relation $P = K_1\rho^{5/3}$ for low electron densities, and $P = K_2\rho^{4/3}$ in the ultrarelativistic limit. A detailed consideration of equilibrium configurations built on

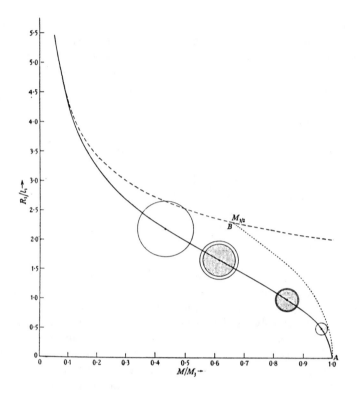

Figure 1.1: The solid line represents the exact mass-radius relation for completely degenerate configurations. The mass, along the abscissa, is measured in units of the limiting mass (denoted by M_3), and the radius, along the ordinate, is measured in the unit $l_1 = 7.72 \times 10^8 \; \mu_e^{-1}$ cm. The dashed curve represents the relation that follows from the equation of state given in equation (1.7); at the point B along this curve, the fermi momentum p_F of the electrons at the center of the configuration is exactly equal to mc. Along the exact curve, at the point where a full circle (with no shaded part) is drawn, p_F (at the center) is again equal to mc; the shaded parts of the other circles represent the regions in these configurations where the electrons may be considered to be relativistic ($p_F \gg mc$). (Reproduced from Chandrasekhar 1935).

the above (exact) equation of state led to a mass-radius relation shown in figure 1.1. While the exact relation (full line) approximates the relation obtained (in the nonrelativistic approximation) for $M \to 0$, it predicted that the *radius tends to zero for* $M \to M_{\text{limit}}$. In other words, *finite degenerate equilibrium configurations exist only for* $M < M_{\text{limit}}$. Given the chemical composition (or equivalently the mean molecular weight per electron μ_e), this limiting mass is uniquely determined by a combination of fundamental constants:

$$M_{\text{limit}} = 0.197 \left[\left(\frac{hc}{G} \right)^{3/2} \cdot \frac{1}{m_H^2} \right] \frac{1}{\mu_e^2}$$

$$= 5.76 \mu_e^{-2} \; M_\odot. \qquad (1.18)$$

This mass has rightly come to be known as the *Chandrasekhar limit*, and plays a central role in the theory of relativistic stars. We shall return to it presently.

1.3　Why are the stars as they are?

After completing his investigations of white dwarfs Chandrasekhar turned to a detailed investigation of the internal constitution of stars. He extended and developed three basic methods. They are

1. The method of integral theorems;

2. The method of homologous transformations;

3. The method of stellar envelopes.

He published the results in a series of very detailed papers, and subsequently distilled and summarized them in his classic monograph *An Introduction to the Study of Stellar Structure*. Here we shall single out one of the integral theorems because it serves to illustrate his approach to the subject as a whole. The method of integral theorems consists in finding *inequalities* for quantities like the central pressure, the mean pressure, the potential energy, the mean value of gravity, etc. The particular theorem we shall refer to concerns the *central pressure* in a star.

But first let us digress a little. As already remarked, Eddington's *standard model* of the stars was enormously successful. Perhaps most importantly, it provided an explanation for the observed masses of stars. It is an extraordinary fact that the overwhelming majority of stars have masses close to that of the Sun: stars with masses very much less than, or very much more than, the mass of the Sun are relatively infrequent. Why is this so? Eddington posed this question to himself and answered it in his parable of a physicist on a cloud-bound planet (*The Internal Constitution of the Stars*):

> The outward flowing radiation may be compared to a wind flowing through the star and helping to distend it against gravity. The formulae to be developed later enable us to calculate what proportion of the weight of the material is borne by this wind, the remainder being supported by the gas pressure. To a first approximation the proportion is the same at all parts of the

star. It does not depend on the density nor on the opacity of
the star. It depends only on the mass and molecular weight.
Moreover, the physical constants employed in the calculation
have all been measured in the laboratory, and no astronomical
data are required. We can imagine a physicist on a cloud-
bound planet who has never heard tell of the stars calculating
the ratio of radiation pressure to gas pressure for a series of
globes of gas of various sizes, starting, say, with a globe of mass
10 g, then 100 g, 1000 g and so on, so that his nth globe contains
10^n g. Table I shows the more interesting part of his results.

Table I

No. of globe	Radiation pressure	Gas pressure	No. of globe	Radiation pressure	Gas pressure
32	0.0016	0.9984	36	0.951	0.049
33	0.106	0.894	37	0.984	0.016
34	0.570	0.430	38	0.9951	0.0049
35	0.850	0.150	39	0.9984	0.0016

The rest of the table would consist mainly of long strings of 9's
and 0's. Just for the particular range of mass about the 33rd to
35th globes the table becomes interesting, and then lapses back
into 9's and 0's again. Regarded as a tussle between matter and
either (gas pressure and radiation pressure) the contest is overwhelmingly one-sided except between numbers 33–35, where we
may expect something to happen.

What *happens* is the stars.

We draw aside the veil of cloud beneath which our physicist has
been working and let him look up at the sky. There he will find
a thousand million globes of gas nearly all of mass between his
33rd and 35th globe—that is to say, between 1/2 and 50 times
the Sun's mass. The lightest known star is about 3×10^{32} g and
the heaviest about 2×10^{35} g. The majority are between 10^{33}
and 10^{34} g, where the serious challenge of radiation pressure to
compete with gas pressure is beginning.

But why is the relative extent to which radiation pressure provides
support against gravity a relevant factor to the "happening" of stars?
A more rational argument, due to Chandrasekhar, is the following. To
quote him, "Domains of natural phenomena are often circumscribed by
well-defined scales, and theories concerning them are successful only to the
extent that these scales emerge naturally in them. Thus, to the question

'why are the atoms as they are?' the answer 'because the Bohr-radius—
$h^2 / (4\pi^2 m_e e^2) \sim 0.5 \times 10^{-8}$ cm—provides a correct measure of their dimensions' is apposite. In a similar vein, we may ask 'why are the stars
as they are?', intending by such a question to seek the basic reason why
modern theories of stellar structure and stellar evolution prevail."

The answer may be found along the following lines. According to one of
the integral theorems proved by Chandrasekhar (1936a, 1936b), the pressure P_c at the center of a star of mass M, in hydrostatic equilibrium and
in which the density $\rho(r)$ at any point r does not exceed the mean density
$\bar{\rho}(r)$ interior to that point r, must satisfy the inequality

$$\frac{1}{2}G\left(\frac{4\pi}{3}\right)^{1/3} \bar{\rho}^{4/3}\ M^{2/3} \leq P_c \leq \frac{1}{2}G\left(\frac{4\pi}{3}\right)^{1/3} \rho_c^{4/3}\ M^{2/3}, \qquad (1.19)$$

where $\bar{\rho}$ denotes the mean density of the star and ρ_c its density at the
center. The meaning of this inequality is the following: the pressure at
the center of a star must be intermediate between those at the centers of
two configurations of *uniform* density, one at a density equal to the mean
density $\bar{\rho}$ and the other at a density equal to the density ρ_c at the center.
If this inequality is violated, then there must be regions in the star where
adverse density gradients prevail, and this will lead to instabilities. Thus,
*satisfying this inequality is a necessary condition for the stable existence of
a star.*

Before proceeding further one must eliminate the explicit temperature
dependence of the central pressure P_c. This is readily done as follows. Let
us introduce the fraction β defined by

$$\begin{aligned}
P &= p_{\text{gas}} + p_{\text{rad}} \\
&= \frac{1}{\beta}p_{\text{gas}} = \frac{1}{1-\beta}p_{\text{rad}} \\
&= \frac{1}{\beta}\frac{\rho kT}{\mu m_H} = \frac{1}{1-\beta}\frac{1}{3}aT^4.
\end{aligned} \qquad (1.20)$$

We may eliminate T from these relations and express the total pressure in
terms of ρ and β. Thus

$$P = \left[\left(\frac{k}{\mu m_H}\right)^4 \cdot \frac{3}{a} \cdot \frac{1-\beta}{\beta^4}\right]^{1/3} \rho^{4/3} \equiv C(\beta)\rho^{4/3}. \qquad (1.21)$$

Let us use this expression for the total pressure in the right-hand part
of the inequality (1.22) as a necessary condition for the existence of a stable
star,

$$\left[\left(\frac{k}{\mu m_H}\right)^4 \cdot \frac{3}{a} \cdot \frac{1-\beta_c}{\beta_c^4}\right]^{1/3} \leq \left(\frac{\pi}{6}\right)^{1/3} GM^{2/3}. \qquad (1.22)$$

Substituting for Stefan's constant a and simplifying, one gets

$$\mu^2 M \left(\frac{\beta_c^4}{1 - \beta_c}\right)^{1/2} \geq 0.19 \left[\left(\frac{hc}{G}\right)^{3/2} \cdot \frac{1}{m_H^2}\right]. \qquad (1.23)$$

We observe that the above inequality has isolated the following combination of fundamental constants of the dimensions of a mass:

$$\left(\frac{hc}{G}\right)^{3/2} \frac{1}{m_H^2} \simeq 29.2 M_\odot, \qquad (1.24)$$

and that this mass is of *stellar magnitude* (Chandrasekhar 1937). This inequality also provides an *upper limit* to $(1 - \beta_c)$ for a star of a given mass:

$$1 - \beta_c \leq 1 - \beta_\star, \qquad (1.25)$$

where $(1 - \beta_\star)$ is uniquely determined by the mass M of the star and the mean molecular weight, μ, by the quartic equation

$$\mu^2 M = 5.48 \left[\frac{1 - \beta_\star}{\beta_\star^4}\right]^{1/2} M_\odot. \qquad (1.26)$$

From this equation it follows that for a star of solar mass (and $\mu = 1$) *the radiation pressure at the centre cannot exceed 3% of the total pressure.* (It also follows that this fraction increases with increasing mass—a point to which we shall return later.) What do we conclude from the foregoing calculation? *"We conclude that to the extent eq. (1.26) is at the base of the equilibrium of actual stars, to that extent the combination of natural constants* $(hc/G)^{3/2} \left(1/m_H^2\right)$, *providing a mass of proper magnitude for the measurement of stellar masses, is at the base of a physical theory of stellar structure"* (Chandrasekhar 1984).

1.4   On the evolution of the main sequence of stars

By the end of the 1930s many of the major questions concerning the structure of the main-sequence stars were settled, and the attention shifted to the evolution of the stars away from the main sequence. In their seminal papers Weizsäcker (1938) and Bethe (1939) had established that the source of energy radiated by the main-sequence stars is the transformation of hydrogen into helium through the "C-N-O cycle." Gamow was among the first to formulate a picture of stellar evolution on the basis of the Bethe-Weizsäcker theory (1939 a,b). His model was based on three assumptions: (i) the stars evolve gradually through a sequence of *equilibrium configurations*; (ii) successive equilibrium configurations are homologous; and (iii) nuclear reactions continue to take place till all the hydrogen in the star is exhausted.

Chandrasekhar turned his attention to this problem around 1940, and two of his papers from that period (Henrich and Chandrasekhar 1941; Schönberg and Chandrasekhar 1942) turned out to be landmark papers. In this section we shall briefly summarize the main conclusions of these two papers, and also attempt to place them in perspective against our present understanding of stellar evolution. Both these papers are devoted to a discussion of stellar models with *isothermal cores*. At the end of hydrogen burning, the star is left with an inert helium core surrounded by a hydrogen-rich envelope. Hydrogen contines to burn in a *shell* at the bottom of the envelope. The helium core must be nearly isothermal, and hence the attempt to construct stellar models with isothermal cores and radiative envelopes. Henrich and Chandrasekhar considered the case in which the value of the mean molecular weight, μ, was the same in both regions. In the subsequent paper, Schönberg and Chandrasekhar discussed the more general (and more relevant) case of different molecular weights μ_e and μ_c, for the envelope and the core, respectively. The basic approach was to require that at the interface the values of the pressure, temperature, and mass of the core should be identical:

$$P(r_i)_{\text{core}} = P(r_i)_e; \quad T(r_i)_{\text{core}} = T(r_i)_e; \quad M(r_i)_{\text{core}} = M(r_i)_e, \quad (1.27)$$

where P, T and M denote the total pressure, the temperature, and the mass within the radius, respectively. The subscript i indicates that the values refer to the interface and the subscript e indicates that the quantities correspond to the envelope solutions of the equilibrium equations. The above conditions are the only ones to be satisfied, and so one gets a family of configurations for any given set of M, R, and L. To fit the core and envelope solutions at the interface they introduced the *homology invariants* U and V to describe the isothermal core,

$$U = \frac{4\pi r^3 \rho_{\text{core}}}{M(r)}, \quad V = \frac{\mu_c m_H}{k} \cdot \frac{GM(r)}{rT(r)}. \quad (1.28)$$

The equations of fit in the new variables are

$$U_i = 4\pi \left[\frac{r^3 \rho(r_i)_e}{M(r_i)_e} \right] \frac{\mu_c}{\mu_e}, \quad V_i = \left[\frac{\mu_e m_H}{k} \cdot \frac{GM(r_i)_e}{r_i T(r_i)_e} \right] \frac{\mu_c}{\mu_e}. \quad (1.29)$$

The method employed in these papers—which then became very widely used in the literature—was to "fit" the core and envelope solutions in the U-V plane by finding their *intersection*. Since the quantities q and ν

$$q = \frac{r_i}{R} \quad \text{and} \quad \nu = \frac{M(r_i)}{M} \quad (1.30)$$

are homology invariants, they could be used to label the different configurations corresponding to the same stellar mass and the same central temperature. Using this technique Chandrasekhar and his colleagues were able

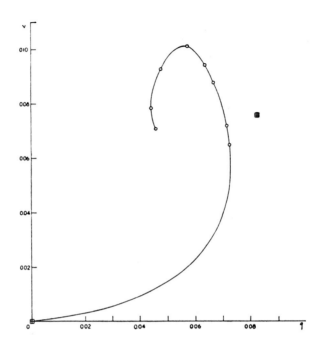

Figure 1.2: Plotted along the abscissa is the radius of the isothermal core as a fraction of the stellar radius ($q = r_i/R$), and along the ordinate is the mass of the core as a fraction of the stellar mass ($\nu = M_c/M$). The circles refer to models with isothermal cores, and the squares refer to models with convective core. Starting from its minimum value, ν increases rapidly as q grows, reaches its absolute maximum, and starts spiraling. The absolute maximum is the Schönberg-Chandrasekhar limiting mass (from Schönberg and Chandrasekhar 1942).

to derive several properties of models with isothermal cores and radiative envelopes. Their most important conclusion—as borne out by subsequent developments—was the following: *There are no equilibrium configurations with the isothermal cores having masses exceeding a critical mass. This upper limit is a decreasing function of μ_c/μ_e.* In the case of equal molecular weights the upper limit for the ratio $\nu = M_c/M$ is ~ 0.35. Schönberg and Chandrasekhar, who considered the more general case, estimated an upper limit for $M_c/M \sim 0.1$. The dependence of the fractional core mass on the fractional core radius is reproduced in figure 1.2. As may be seen, there are no equilibrium configurations with cores containing less than 0.065 or more than 0.1 of the stellar mass. The lower limit is due to the appearance of convective instability at the interface, while the upper one is due to the impossibility of fitting a core to an envelope. Starting from its minimum value, ν increases rapidly as q grows, reaches a maximum, and starts spi-

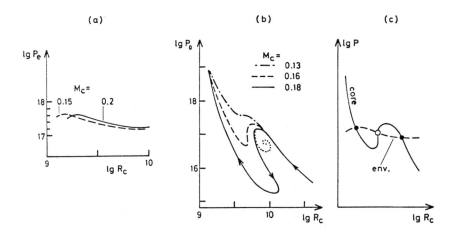

Figure 1.3: Models with isothermal cores and radiative envelopes. (a) The pressure P_e at the bottom of the envelope (in dyn cm^{-2}) plotted against the core radius R_c for a $2M_\odot$ star. The two curves correspond to two values of the core mass (in $M_\odot$). Thus the envelope solution for the interface is nearly independent of M_c. (b) The pressure P_0 at the surface of the core for different core masses (in $M_\odot$). The arrows along the solid curve indicate the direction of increasing central pressure. The curve spirals around a certain value (the dotted curve) if degeneracy of the core is neglected. The inclusion of degeneracy unwinds the spiral, and it rises once again for increasing central pressure. (c) The intersection of the core and envelope solutions. The filled circles represent stable solutions, and the open circle represents an unstable solution (from computations by Roth (1973) reproduced from Kippenhahn and Weigert 1990).

raling around a certain value (not shown in the figure reproduced here but discussed in the text).

This upper limit to the mass of the isothermal core has come to be known as the *Schönberg-Chandrasekhar limit*. This limit is certainly exceeded by helium cores left behind after central hydrogen burning in stars of the upper main sequence. What, then, is the significance of this limit? This became clear only after detailed numerical results (which also included the degeneracy of the core) became available. In view of the fundamental significance of this result, we shall summarize our present understanding of this problem.

For illustration we show the results of detailed numerical integration by Roth (1973) which has been reproduced from the excellent modern textbook on the subject by Kippenhahn and Weigert (1990). Figure 1.3(a) shows the pressure at the bottom of the envelope as a function of the core radius for two assumed values for the core mass. As would be expected, the pressure is insensitive to the core mass. Figure 1.3(b) shows the behavior of the pressure at the surface of the isothermal core for several values of the core mass. Consider first the solid curve, which is for $M_c = 0.18M_\odot$ and $T_0 = 2.24 \times 10^{7\circ}$K. If the degeneracy of the core is neglected, then with increasing central pressure this curve would rise from the lower right hand part of the diagram and spiral in as indicated by the *dotted* curve. The inclusion of degeneracy "unwinds" this spiral and it begins to rise again (degeneracy becomes more and more important with increasing central pressure). As may be seen, the minimum is less and less pronounced for smaller core masses. Next let us consider the intersection of the core and envelope solutions. For very small values of the core mass, the core solution is monotonic and there is only one intersection. When the mass of the core increases to a critical value—the Schönberg-Chandrasekhar limit—the core solution develops a pronounced maximum, and if the envelope solution passes between the maximum and minimum there will be three intersections: the one with the largest R_c corresponds to the core described by an ideal gas, the one with intermediate R_c corresponds to partial degeneracy, and the one with the smallest R_c corresponds to large degeneracy of the core.

The emerging picture is best summarized by figure 1.4 in which linear series of complete equilibrium solutions are shown for stars with four different masses (in units of $M_\odot$). R_c is the radius and $M_c = q_c M$ is the mass of the isothermal core. The curves for the more massive stars have three branches; the solid sections represent thermally stable branches and the dashed section represents unstable models. On the upper branch the cores are nondegenerate, but they are strongly degenerate in the lower branch. *The turning point with the larger value of core mass* (q_{SC}) *defines the Schönberg-Chandrasekhar limit.* As we go to smaller stellar masses the two turning points approach each other, and at $M \sim 1.4M_\odot$ they merge and disappear.

Let us conclude this discussion of the Schönberg-Chandrasekhar limit with a brief discussion of its implication for the evolution of the stars in the upper main sequence. To be specific, let us consider a 3 solar mass star and follow its evolution with the aid of figure 1.5(a). When the mass of the isothermal helium core is still relatively small, the star will settle into an equilibrium state represented by the upper branch. This corresponds to a dwarf star close to the main sequence (as shown in fig. 1.5b). As the mass of the helium core grows due to hydrogen burning in the shell, the star will "move" along this upper branch, maintaining equilibrium. This proceeds continuously till the core mass reaches the Schönberg-Chandrasekhar limit. When the core mass exceeds this critical value the only equilibrium models are in the lower branch and the core will have to contract discontinuously. This contraction of the core will be accompanied by an expansion of the star and the star will move rapidly in the HR diagram from the main sequence

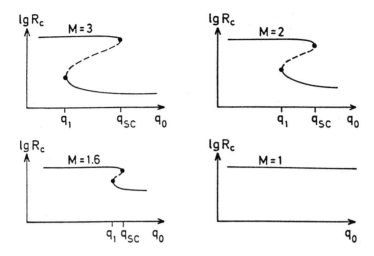

Figure 1.4: Linear series of complete equilibrium solutions for four different stellar masses M (in $M_\odot$) having isothermal core of mass $M_c = q_0 M$. Each solution is characterized by its core radius R_c and its relative core mass q_0. Branches with thermally stable solutions are shown by solid lines, and the unstable branches are shown as dashed lines. The turning point $q_0 = q_{SC}$ defines the Schönberg-Chandrasekhar limit. (After Roth 1973; from Kippenhahn and Weigert 1990).

to the region of the Hayashi line. This central conclusion, namely that the core will contract and the star will transform itself into a red giant, is borne out by a detailed evolutionary calculation such as the one shown in figure 1.6. The core contraction (and the expansion of the star) occurs roughly on the Kelvin-Helmholtz timescale of the core ($\sim 3 \times 10^6$ yr for a $5M_\odot$ star). In contemporary literature this consequence of the Schönberg-Chandrasekhar limit is taken as the explanation for the existence of the well known *Hertzsprung gap* in the HR diagram, a region between the main sequence and the red giants where there is a paucity of observed stars.

Such a sudden contraction of the core may also have implications for dramatic mass loss from the upper main-sequence stars. The discovery of white dwarfs in some open clusters with main-sequence turn-off mass $\sim 6M_\odot$ certainly implies such a mass loss. For otherwise, according to our current understanding, the degenerate carbon cores of stars with mass $\geq 6M_\odot$ will eventually grow to $\sim 1.4M_\odot$ and ignite, resulting in a detonation of the star as a Type I supernova. Whether the Schönberg-Chandrasekhar limit has anything to do with dramatic mass loss from the stars of the upper main sequence or not, we would like to conclude this section by quoting from the last paragraph of this fundamental paper:

It therefore appears difficult to escape the conclusion that be-

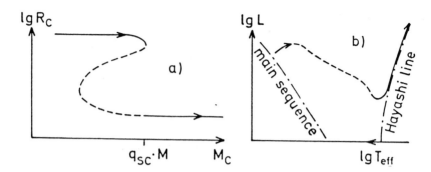

Figure 1.5: (a) Linear series of equilibrium solutions for a 3 solar mass star. With increasing core mass, the model shifts along the solid lines as indicated by the arrows. When the mass of the core grows to $q_{SC}M$—*the Schönberg-Chandrasekhar critical mass*—the core will have to contract discontinuously. This results in an expansion of the star, and the star moves rapidly from near the main sequence to the red giant branch in the HR diagram. The rapid transit of the star across the *Hertzsprung gap* is indicated by the dashed line in (b) (from Kippenhahn and Weigert 1990).

yond this point the star must evolve through nonequilibrium configurations. It is difficult to visualize what form these non-equilibrium transformations will take; but, whatever their precise nature, they must depend critically on whether the mass of the star is greater or less than the upper limit M_3 $(= 5.7\mu^{-2}M_\odot)$ to the mass of degenerate configurations. For masses less than M_3 the nonequilibrium transformations need not take particularly violent forms, as finite degenerate white-dwarf states exist for these stars. However, when $M > M_3$, the star must eject the excess mass first, before it can evolve through a sequence of composite models consisting of degenerate cores and gaseous envelopes toward the completely degenerate state. *Our present conclusions tend to confirm a suggestion made by one of us (S.C.) on different occasions that the supernova phenomenon may result from the inability of a star of mass greater than M_3 to settle down to the final state of complete degeneracy without getting rid of the excess mass.*

1.5   The fate of massive stars

Let us now return to the early 1930s once again. The discovery of the limiting mass for white dwarfs in which the gravitational pressure is balanced by the degeneracy pressure of the electrons was to play a central

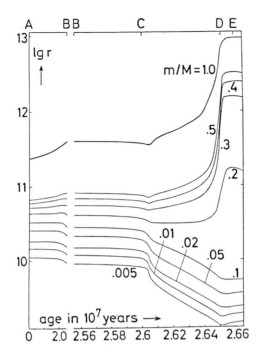

Figure 1.6: The radial variations of different mass shells characterized by their m/M values in the post-main-sequence phase of a $7M_\odot$ star. The rapid contraction of the core and the expansion of the envelope is clearly seen (from Hofmeister *et al.* 1964).

role in astrophysics in the second half of this century. Equally fundamental and far-reaching in its significance was a paper Chandrasekhar published in 1932, entitled *"Some remarks on the state of matter in the interior of stars."*

Although the astronomical community was indifferent to the isolation of the limiting mass of white dwarfs, and its high priests hostile to the idea, Chandrasekhar took the result seriously and attempted to relate it to the life history of gaseous stars with masses greater than the critical mass. The first question to be resolved was the condition under which a star, initially gaseous, can develop a degenerate core. This question could be answered by comparing the electron pressure as given by the classical perfect gas equation of state with the expression for the degeneracy pressure (see fig. 1.7). Since the former is a function of both ρ and T, while the latter is independent of temperature, Chandrasekhar expressed the classical pressure in terms of ρ and β_e, where the fraction of β_e is defined through the relation

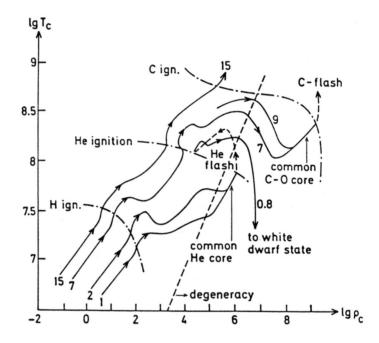

Figure 1.7: This plot of $\log p$ *vs.* $\log \rho$ illustrates the onset of degeneracy for increasing density at constant β. The straight line ABK represents the equation of state $p = K_1 \rho^{5/3}$ and BC the equation of state $p = K_2 \rho^{4/3}$. ABC gives roughly the equation of state of a degenerate gas. DE represents the classical equation of state (eq. (1.32) with $\mu = 2$ and $\beta = 0.98$). It intersects the degenerate equation of state (AB, C) at E. Thus for a star with $\beta = 0.98$ there are two surfaces of demarcation: a perfect gas envelope, a degenerate zone EB, and then a relativistically degenerate zone. If $\beta = 0.9079$, then GB represents the perfect gas equation of state and the degenerate zone reduces to a single layer, and the relativistically degenerate zone is described equally well by the perfect gas equation. Now if $\beta < 0.9079$, the perfect gas equation of state has *no* intersections with ABC, and this means that however high the density may become, the temperature rises sufficiently rapidly to prevent the matter from becoming degenerate (from Chandrasekhar 1932).

$$P_{\text{tot}} = \frac{1}{\beta_e}\left(\frac{k}{\mu_e m_H}\rho T\right) = \frac{1}{1-\beta_e}\left(\frac{1}{3}aT^4\right). \qquad (1.31)$$

One gets

$$p_e = \left[\left(\frac{k}{\mu_e m_H}\right)^4 \cdot \frac{3}{a} \cdot \frac{1-\beta_e}{\beta_e}\right]^{1/3}\rho^{4/3}. \qquad (1.32)$$

If degeneracy sets in at all in the core it will do so under conditions in which special relativistic corrections would be important. Hence the above expression should be compared with the *relativistic degeneracy pressure*

$$p_{\text{deg}} = K_2 \rho^{4/3}, \quad K_2 = \frac{1}{8} \left(\frac{3}{\pi}\right)^{1/3} \frac{hc}{(\mu_e m_H)^{4/3}}. \tag{1.33}$$

As Chandrasekhar pointed out, *in this connection it will have to be remembered that considerations of relativity do not affect the equation of state of a perfect gas. $p = NkT$ is true independent of relativity!* Thus, if

$$\left[\left(\frac{k}{\mu_e m_H}\right)^4 \cdot \frac{3}{a} \cdot \frac{1 - \beta_e}{\beta_e}\right]^{1/3} > K_2, \tag{1.34}$$

then the pressure p_e given by the classical equation of state will be greater than the degeneracy pressure, *not only for the prescribed ρ and T, but for all ρ and T having the same β_e.* Inserting the value for Stefan's constant a, the above inequality reduces to

$$\frac{960}{\pi^4} \cdot \frac{1 - \beta_e}{\beta_e} > 1 \tag{1.35}$$

or, equivalently,

$$1 - \beta_e > 0.0921 = 1 - \beta_\omega \ (\text{say}). \tag{1.36}$$

Thus, *the criterion for a star to develop degeneracy is that the radiation pressure be less than 9.2% of the total pressure.* This exact result is of singular importance in all the contemporary schemes of stellar evolution.

It is an important result of the *standard model* due to Eddington that radiation pressure must play a more dominant role as the mass of a star increases. Chandrasekhar proceeded to calculate the mass of a star (in the standard model) in which radiation pressure is precisely equal to 9.2% of the total pressure. It may be recalled that in the standard model the fraction β is assumed to be constant throughout the star. Under this assumption, stars are polytropes of index 3, since the total pressure can be written as

$$P = \left[\left(\frac{k}{\mu m_H}\right)^4 \cdot \frac{3}{a} \cdot \frac{1 - \beta}{\beta^4}\right]^{1/3} \rho^{4/3} = C(\beta)\rho^{4/3}. \tag{1.37}$$

For such a star the mass is uniquely determined by the constant of proportionality in the polytropic equation of state:

$$M = 4\pi \left(\frac{C(\beta)}{\pi G}\right)^{3/2} \times 6.89. \tag{1.38}$$

For $\beta = \beta_\omega = 0.908$ this gives

$$M = 6.65 \mu^{-2} M_\odot. \tag{1.39}$$

Thus, in the standard model, stars with masses exceeding $6.65\mu^{-2}M_\odot$ will have radiation pressure that will exceed 9.2% of the total pressure, and consequently they cannot, during the course of their evolution, develop degeneracy in their interiors, and, accordingly, an eventual white-dwarf state is impossible for them without a substantial ejection of mass. Although this remarkable conclusion was soundly rejected by Eddington and Milne—two of the most distinguished and influential astrophysicists of that time—Chandrasekhar himself was so convinced of his result that he asserted with supreme confidence:

> For all stars of mass greater than $6.6\mu^{-2}M_\odot$, the perfect gas equation of state does not break down, however high the density may become, and the matter does not become degenerate. An appeal to Fermi-Dirac statistics to avoid the central singularity cannot be made.

Although convinced of it, Chandrasekhar was nevertheless uneasy about the above conclusion. Since infinite density cannot be entertained, and since no other equation of state was available at that stage, he invoked the assumption that there must exist a *maximum density* ρ_{max} which matter is capable of. Accordingly he constructed models with gaseous envelopes and homogeneous cores at the maximum density of matter (at the density of nuclear matter). But he concluded this remarkable paper on a cautious note:

> Great progress in the analysis of stellar structure is not possible before we can answer the following fundamental question: *Given an enclosure containing electrons and atomic nuclei (total charge zero) what happens if we go on compressing the material indefinitely?*

It should be remarked that in 1932 (when the paper discussed above was published) he had not yet convinced himself of the significance of the maximum mass of white dwarfs. However, after working out the exact theory of white dwarfs, he concluded (Chandrasekhar 1934a):

> Finally, it is necessary to emphasize one major result of the whole investigation, namely, that it must be taken as well established that the life-history of a star of small mass must be essentially different from the life-history of a star of large mass. For a star of small mass the natural white-dwarf stage is an initial step towards complete extinction. A star of large mass cannot pass into the white-dwarf stage, and one is left speculating on other possibilities.

Uncharacteristically, he even speculated (Chandrasekhar 1934b):

It is conceivable, for instance, that at very high critical density the atomic nuclei come so near one another that the nature of the interaction might suddenly change and be followed subsequently by a sharp alteration in the equation of state in the sense of giving a maximum density of which matter is capable. However, we are now entering a region of pure speculation, and it is best to conclude the discussion at this stage.

But in a paper published around the same time Baade and Zwicky (1934) were less cautious:

With all reserve we advance the view that supernovae represent the transitions from ordinary stars into neutron stars, which in their final stages consist of extremely closely packed neutrons.

In a prescient paper published in 1939, Oppenheimer and Volkoff pointed out that as one approached the limiting mass of white dwarfs along the sequence of the completely degenerate configurations, the central density will become high enough for the electrons at the fermi level to combine with the protons to form neutrons. Thus beyond a critical density neutrons will be the more stable particles. Oppenheimer and Volkoff studied the mass-radius relation of neutron stars with the aid of general relativistic equations for hydrostatic equilibrium. Based on this work Chandrasekhar (1939a) concluded the following regarding the fate of stars more massive than the limiting mass of white dwarfs, but less massive than $6.65\mu^{-2}M_\odot$ (in the standard model):

If the degenerate cores attain sufficiently high densities (as is possible for these stars) the protons and electrons will combine to form neutrons. This would cause a sudden diminution of pressure resulting in the collapse of the star onto a neutron core giving rise to an enormous liberation of gravitational energy. This may be the origin of the supernova phenomenon.

Those, then, were the predictions made nearly sixty years ago. Have they been confirmed by observations, as well as detailed computations? Yes! As may be seen in figure 1.8, a star of, say, $15M_\odot$ does not develop degeneracy in the core during successive stages of nuclear burning. The fusion reactions proceed in a controlled fashion till an iron core forms. And degeneracy finally sets in in the iron core for the following reason. In stars with masses $\geq 10M_\odot$, radiation pressure always remains in excess of 9.2% of the total pressure (i.e., $1 - \beta_e > 1 - \beta_w$). However, when the carbon core finally ignites, there will be a copious emission of neutrinos resulting in a cooling of the core, thus lowering $(1 - \beta_e)$; but it will still be in excess of $(1 - \beta_w)$. The increase in density and temperature will eventually result in neon ignition. The resultant neutrino emission will further lower $(1 - \beta_e)$. It is

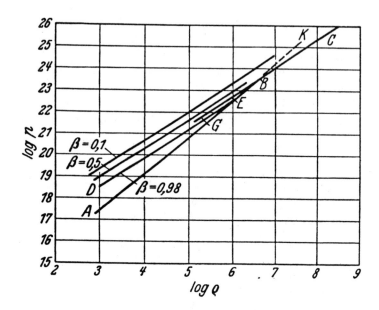

Figure 1.8: The evolution of the central temperature T_c (in °K) and central density ρ_c (in g cm^{-3}) for stars of different masses. The conditions for the ignition of hydrogen, helium, and carbon are indicated by dot-dashed lines. In the region to the left of the sloping (dashed) straight line, the core of the star is described by the classical perfect gas equation of state. As may be seen, the core of a $15M_\odot$ star remains nondegenerate through successive stages of nuclear ignition. This is because in such massive stars the radiation pressure exceeds 9.2% of the total pressure. Beyond the carbon burning phase the large neutrino luminosity lowers radiation pressure selectively in the central region, and eventually it becomes less than 9.2% of the total pressure. Consequently relativistic degeneracy sets in, in the iron core (from Iben 1974).

important to appreciate that the emission of neutrinos occurs selectively in the central region. This succession of nuclear ignitions followed by a lowering of $(1 - \beta_e)$ will continue till $(1 - \beta_e)$ becomes *less* than $(1 - \beta_\omega)$, and a *relativistically degenerate iron core will form*. The mass of this core will quickly grow to $\sim 1.4M_\odot$. Soon instability of some sort will set in, resulting in the collapse of the core. Let us examine this more closely. In the framework of the Newtonian theory of gravitation, for the spherical core to be stable against radial perturbations the pressure-averaged value of the adiabatic index Γ must be greater than 4/3. That is,

$$\bar{\Gamma} = \frac{\int_0^M \Gamma(r)P(r)\, dM(r)}{\int_0^M P(r)\, dM(r)} > \frac{4}{3}, \qquad (1.40)$$

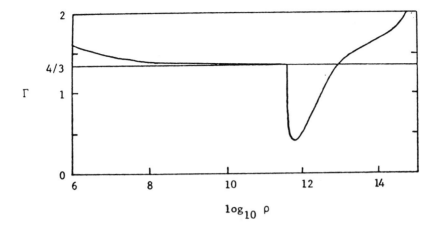

Figure 1.9: The adiabatic index as a function of the mass density. In Newtonian physics this index has to be greater than 4/3 for stability. Inclusion of general relativistic effects increases the critical value to above 4/3.

where

$$\Gamma = \frac{\Delta P/P}{\Delta \rho/\rho}. \qquad (1.41)$$

If $\bar{\Gamma} < 4/3$, *dynamical instability* of a global character will set in, with a characteristic timescale given by the sound travel time across the core. Figure 1.9 schematically shows the behavior of Γ as a function of density. As the density approaches $\sim 10^9$ g cm^{-3}, Γ will tend to 4/3 and matter will become marginally stable. When the density reaches $\sim 10^{10}$ g cm^{-3}, inverse β-decay will set in, resulting in the neutronization of matter. The decrease in the number density of electrons will cause a diminution of the pressure, resulting in the collapse of the core. Initially the number of free neutrons will be small, and consequently they will not contribute significantly to the pressure. Eventually, when the density increases to $\sim 7 \times 10^{12}$ g cm^{-3}, the degeneracy pressure of the free neutrons will become significant and the matter will become stable once again. The *mean density* of the collapsed core (corresponding to $\bar{\Gamma} > 4/3$) will be $\sim 10^{14}$ g cm^{-3}. The resultant configuration will be a *neutron star* of mass $\sim 1.4 M_\odot$.

As already mentioned, the above considerations were within the framework of the Newtonian theory of gravitation. As we saw, in this picture, neutronization of matter played a central role in the instability setting in. The situation is qualitatively different when one examines the same problem in the framework of the general theory of relativity. This discovery was made by Chandrasekhar in 1964 (and is discussed by J. Friedman in this volume). Chandrasekhar showed that in the post-Newtonian approximation to the general theory of relativity, the instability for radial perturbations will set in for all stars with

$$R < \frac{K}{\Gamma - 4/3} \cdot \frac{2GM}{c^2}, \qquad (1.42)$$

where K is a constant. Chandrasekhar and Tooper (1964) applied this result for degenerate configurations near the limiting mass. Since the electrons in these highly relativistic configurations have velocities close to that of light, the effective value of Γ will be close to 4/3. Thus *the post-Newtonian instability will set in for a mass slightly less than the limiting mass* because the modified stability criterion requires

$$\Gamma > \frac{4}{3} + K \frac{2GM}{Rc^2}. \qquad (1.43)$$

The radius of the configuration when this global instability sets in will be $\sim 5 \times 10^3 R_S$, where $R_S = 2GM/c^2$.

1.6   Epilogue

The burst of neutrinos from the 1987a supernova in the Large Magellanic Cloud, the discovery of pulsars in the Crab Nebula and other supernova remnants, and the fact that the measured masses of neutron stars are almost equal to $1.4 M_\odot$ are spectacular confirmations of the remarkable predictions made by Chandrasekhar in the 1930s.

Will all massive stars find peace as neutron stars? Since there is a maximum mass for neutron stars (in analogy with the Chandrasekhar mass limit for white dwarfs) the answer must be "no". Thus in sufficiently massive stars "*an appeal to the Fermi-Dirac statistics to avoid the central singularity cannot be made.*" Eddington clearly recognized the significance of this result. He thus stated (Eddington 1935):

> The star apparently has to go on radiating and radiating and contracting and contracting until, I suppose, it gets down to a few kilometres radius when gravity becomes strong enough to hold the radiation and the star can at last find peace.

But he denied the existence of an upper limit to mass of completely degenerate configurations (white dwarfs and neutron stars), and consequently rejected the above possibility. In a paper published in 1939, Oppenheimer and Snyder were unequivocal about the fate of sufficiently massive stars:

> When all thermonuclear sources of energy are exhausted a sufficiently heavy star will collapse. This contraction will continue

indefinitely till the radius of the star approaches asymptotically its gravitational radius. Light from the surface of the star will be progressively reddened and can escape over a progressively narrower range of angles till eventually the star tends to close itself off from any communication with a distant observer. Only its gravitational field persists.

In modern terminology, it will become a *black hole*.

Chandrasekhar began his research with a detailed study of white dwarfs. The results he obtained in the 1930s are at the base of much of relativistic astrophysics. In particular, his study led inescapably to the conclusion that sufficiently massive stars will ultimately find peace as black holes. He returned to this subject in 1964, and devoted the next thirty years to a detailed study of black holes and singularities.

Chandrasekhar credited Eddington with the founding of modern theoretical astrophysics, and creating the discipline of the structure, the constitution, and the evolution of the stars. The second half of this century will be remembered as the golden age of relativistic astrophysics. This is the era of neutron stars and black holes. Chandrasekhar said the first words on these, and went on to erect major pillars on which the superstructure of contemporary astrophysics rests.

References

Anderson, W. 1929, *Zeits. für Phys.*, **56**, 851.

Baade, W., Zwicky, F. 1934, *Phys. Rev.*, **45**, 138.

Bethe, H. 1939, *Phys. Rev.*, **55**, 434.

Chandrasekhar, S. 1928, *Indian Journal of Physics*, **3**, 241.

Chandrasekhar, S. 1929, *Proc. R. Soc.*, **A125**, 231.

Chandrasekhar, S. 1931a, *Phil. Mag.*, **11**, 592.

Chandrasekhar, S. 1931b, *Astrophys. J.*, **74**, 81.

Chandrasekhar, S. 1931c, *Mon. Not. R. Astr. Soc.*, **91**, 456.

Chandrasekhar, S. 1932, *Z. Astrophys.*, **5**, 321.

Chandrasekhar, S. 1934a, *Observatory*, **57**, 373.

Chandrasekhar, S. 1934b, *Observatory*, **57**, 93.

Chandrasekhar, S. 1935, *Mon. Not. R. Astr. Soc.*, **95**, 207.

Chandrasekhar, S. 1936a, *Mon. Not. R. Astr. Soc.*, **96**, 644.

Chandrasekhar, S. 1936b, *Observatory*, **59**, 47.

Chandrasekhar, S. 1937, *Nature*, **139**, 757.

Chandrasekhar, S. 1939, *An Introduction to the Study of Stellar Structure* (Chicago: University of Chicago Press).

Chandrasekhar, S. 1939a, *Colloque International d'Astrophysics XIII; in Novae and White Dwarfs* (Paris: Hermann and Cie, 1941).

Chandrasekhar, S., Tooper, R. F. 1964, *Astrophys. J.*, **139**, 1396.

Chandrasekhar, S. 1984, *Rev. Mod. Phys.*, **56** (2), 137.

Eddington, A. S. 1926, *The Internal Constitution of Stars* (Cambridge: Cambridge University Press).

Eddington, A. S. 1935, *Observatory*, **58**, 37.

Fowler, R. H. 1926, *Mon. Not. R. Astr. Soc.*, **87**, 114.

Gamow, G. 1939a, *Phys. Rev.*, **55**, 718.

Gamow, G. 1939b, *Nature*, **144**, 575.

Henrich, L. R., Chandrasekhar, S. 1941, *Astrophys. J.*, **94**, 525.

Hofmeister, E., Kippenhahn, R., Weigert, A. 1964, *Zeits. Astrophys.*, **59**, 242.

Iben, I. Jr., 1974, *Ann. Rev. Astron. Astrophys.*, **12**, 215.

Kippenhahn, R., Weigert, A. 1990, *Stellar Structure and Evolution* (New York: Springer-Verlag).

Oppenheimer, J. R., Volkoff, G. 1939, *Phys. Rev.*, **55**, 374.

Oppenheimer, J. R., Snyder, H. 1939, *Phys. Rev.*, **56**, 455.

Roth, M. L. 1973, dissertation (University of Hamburg).

Schönberg, M., Chandrasekhar, S. 1942, *Astrophys. J.*, **96**, 161.

Stoner, E. C. 1929, *Phil. Mag.*, **7**, 63.

Stoner, E. C. 1930, *Phil. Mag.*, **9**, 944.

Weizsäcker, C. F. von. 1938, *Phys. Zs.*, **39**, 633.

2

Neutron Stars Before 1967 and My Debt to Chandra

E. E. Salpeter

At least in his later years, Chandra was particularly famous for general relativity, and throughout his brilliant career he was a model of mathematical rigor and elegance. I have never had a strong interest in general relativity, I am mathematically about as sloppy as one can get away with, and I have spent little time in Chicago. Because of this orthogonality, I have probably had less overlap with Chandrasekhar than most theoretical astrophysicists, and yet even in my case he has had a strong influence. I will illustrate this with a purely personal essay on my own work on equations of state and compact objects, especially neutron stars.

It is interesting to speculate on why some topics are studied when they are, and I have put "neutron stars before 1967" in the title, because the reasons for "why" are clear after 1967: pulsars were discovered (Hewish *et al.* 1968), it became clear that they are rotating neutron stars (Gold 1968), and radiation mechanisms were discussed even just before the discovery (Pacini 1967). In the 40 years before this, on the other hand, there were few practical reasons to study neutron stars, except for the prescient suggestion of neutron stars in supernova remnants (Baade and Zwicky 1934). When I was a graduate student in the 1940s I was unaware of this paper, but my interest was aroused in a very indirect way by the earlier controversy between Chandrasekhar and Eddington on the equation of state for relativistic white dwarf stars. In astrophysics circles this controversy is usually described in terms of Eddington as a great man with deep philosophical beliefs and unorthodox views on how the laws of science might change—i.e., it was not clear whether he was morally right in "putting down" a young man so thoroughly and consistently, but it was not clear either till much later that he was scientifically wrong. However, in 1946 I was a graduate student in physics, not in astrophysics, my thesis advisor was Rudolf Peierls, and it *was* clear that Eddington was wrong right from the start! At least this was the situation with two very specific papers of Eddington's.

These two papers (Eddington 1935a,b) were mainly concerned with the laws of physics in existence at the time, especially quantum mechanics and special relativity, not with philosophy or the future (in one of them there was one delightful digression into the "magic numbers" in astronomy and physics which was vintage Eddington, but this did not impinge on the main text). There were two aspects to these papers: (i) they pointed out genuine difficulties that would be faced if one wanted to carry out very rigorous and very accurate calculations, and (ii) they presented an explicit calculation of the equation of state for relativistic electrons as Fermi-Dirac particles which not only gave the wrong result but consisted of sheer nonsense or double-talk or both! An example of (i) was how to treat Dirac electrons under high pressure, when they are not free particles but are confined by a strong gravitational field. My thesis advisor had solved this problem within a year (Peierls 1936), although it was not a trivially simple calculation. And I have worried off and on over the last 50 years about (ii). Eddington was a great man and on some level of consciousness he must have known he had written nonsense—how could he live with himself and how could two respectable journals publish such papers? I have felt that much of the answer stems from the genuine problems in (i) obscuring the treatment in (ii). I consider the juxtaposition of macroscopic and several microscopic complications in one problem a particularly exciting challenge for a theorist.

Some of the questions raised in the two Eddington papers had to do with interactions between particles, directly and through Coulomb forces, i.e., forerunner questions for the combination of plasma physics and quantum mechanics. I have worked on this combination off and on since then, stimulated not only by the negative influence of the two Eddington papers, but also by the positive influence of Chandra's numerous papers in the 1930s on the equation of state and white dwarf star structure. These papers (e.g., Chandrasekhar 1935), and my thesis advisor's paper on Dirac electrons in a large-scale potential field, actually were not easy reading and required appreciable effort on the part of a young and inexperienced graduate student to absorb. However, they were so methodical, detailed, and logically constructed that, once absorbed, they acted as models for how even a youngster could write papers in the future. To digress on contrasting styles—Landau (and, in other areas, Fermi) had written brilliant papers which seemed to be easy reading at first sight but were not easy to use as role models for common mortals. Oppenheimer was smart enough to use Landau's classic paper (Landau 1932) as the starting point for his own work on neutron stars (Oppenheimer and Serber 1938; Oppenheimer and Volkoff 1939), but I would not have been. This difference in scientific styles might also be the reason why Landau and Oppenheimer gave so little credit in their papers to Chandra's classic white dwarf papers (Chandrasekhar 1931a,b).

My own first foray into equations of state was not really related to either white dwarfs or neutron stars, but to the plasma physics that goes into the electron screening for thermonuclear reactions (Salpeter 1954). Although I have not discussed this point directly with Schatzman, chapter 4 in his *White Dwarf* book (Schatzman 1958) suggests that he also had been drawn into plasma physics by the Chandra-Eddington controversy on particle interactions (and he worked on electron screening even earlier than I did). Given the absence of any neutron star observations, there was a surprisingly large amount of activity on neutron matter and its equation of state (e.g., Harrison *et al.* 1958; Cameron 1959; Salpeter 1960; to name just a few). This work started to blur the division between white dwarfs and neutron stars or, rather, it provided a region of instability at intermediate densities. More specifically, inverse beta-decays change the charge of nuclei and lead to a maximum white dwarf mass occurring at finite rather than infinite density (e.g., Hamada and Salpeter 1961). Chandrasekhar and Tooper (1964) then showed that general relativity would also have given instability above a finite density even if nuclei were unchangeable (a similar suggestion had already been made in an earlier paper (Kaplan 1949), which was missed by most of us in the west). There was also a brief flurry of activity on neutron stars before the first Dallas Relativity Symposium in December 1963, just in case quasars turned out to be neutron stars, but this false alarm was soon laid to rest. More details will be found in Harrison *et al.* (1965) and in Shapiro and Teukolsky (1983).

My own interest in neutron stars waned somewhat even before neutron stars became a known reality, but not my interest in studying multiple problems, stimulated by Chandra's example of working in many different fields. Three of his many books, on three very different topics, had already appeared well before 1960 and he was well on the way to combining relativity and astrophysics into a new science of relativistic astrophysics. My own excursions into plasma physics plus ionosphere, solid state physics plus interstellar molecules, or accretion flows plus black holes all were pale imitations of this, but all were helped by Chandra's books and by his example.

References

Baade, W., Zwicky, F. 1934, *Phys. Rev.*, **45**, 138.

Cameron, A. G. W. 1959, *Ap. J.*, **130**, 884.

Chandrasekhar, S. 1931a, *Phil. Mag.*, **11**, 592.

Chandrasekhar, S. 1931b, *Ap. J.*, **74**, 81.

Chandrasekhar, S. 1935, *Mon. Not. R. Astr. Soc.*, **95**, 225 and 676.

Chandrasekhar, S., Tooper, R. F. 1964, *Ap. J.*, **139**, 1396.

Eddington, A. 1935a, *Mon. Not. R. Astro. Soc.*, **95**, 194.

Eddington, A. 1935b, *Proc. R. Soc. London*, **152**, 253.

Gold, T. 1968, *Nature*, **218**, 731.

Hamada, T., Salpeter, E. E. 1961, *Ap. J.*, **134**, 683.

Harrison, B. K., Wakano, M., Wheeler, J. A. 1958, in *Onz. Cons. de Physique Solvay* (Brussels: Stoops), 124.

Harrison, B. K., Thorne, K. S., Wakano, M., Wheeler, J. A. 1965, *Gravitation Theory and Gravitational Collapse* (Chicago: University of Chicago Press).

Hewish, A., Bell, S. J., *et al.* 1968, *Nature*, **217**, 709.

Kaplan, S. A. 1949, *Mem. Univ. LWOW*, **15**, 101.

Landau, L. D. 1932, *Phys. Z. Sowjetunion*, **1**, 285.

Oppenheimer, J. R., Serber, R. 1938, *Phys. Rev.*, **54**, 540.

Oppenheimer, J. R., Volkoff, G. M. 1939, *Phys. Rev.*, **55**, 374.

Pacini, F. 1967, *Nature*, **216**, 567.

Peierls, R. 1936, *Mon. Not. R. Astr. Soc.*, **96**, 780.

Salpeter, E. E. 1954, *Austral. J. Phys.*, **7**, 373.

Salpeter, E. E. 1960, *Ann. of Phys.*, **11**, 393.

Schatzman, E. 1958, *White Dwarfs* (Amsterdam: North-Holland).

Shapiro, S. L., Teukolsky, S. A. 1983, *Black Holes, White Dwarfs and Neutron Stars* (New York: Wiley).

3

The Stellar-Dynamical Oeuvre

James Binney

3.1 Introduction

Chandrasekhar was active in stellar dynamics only during the five years 1939–1944. Over this period the focus of his attention varied systematically, so that a chronological ordering of his papers corresponds fairly exactly with an ordering by topic. In sections 3.2 and 3.3 I review in chronological order all the stellar-dynamical papers that he published in refereed journals. Section 3.4 summarizes the content of his book *The Principles of Stellar Dynamics*. Section 3.5 attempts to assess the impact of his writings on the further development of stellar dynamics.

3.2 The ellipsoidal hypothesis

Chandrasekhar entered the field of stellar dynamics with two monumental papers on the ellipsoidal hypothesis. The first paper [1] (154 pages of the *Astrophysical Journal*) dealt with steady-state models, while the second paper (202 pages of the *Astrophysical Journal*) extended the theory to time-dependent models.

The fundamental equation of the theory of collisionless stellar systems is the collisionless Boltzmann equation

$$\frac{\partial f}{\partial t} + \mathbf{v} \cdot \frac{\partial f}{\partial \mathbf{x}} - \frac{\partial \Phi}{\partial \mathbf{x}} \cdot \frac{\partial f}{\partial \mathbf{v}} = 0. \tag{3.1}$$

Eddington (1915) first sought solutions to this equation in which the distribution function f depends on the stellar coordinates $\mathbf{x}$ and velocities $\mathbf{v}$ only through the function

$$Q(\mathbf{x}, \mathbf{v}) \equiv \mathbf{v}' \cdot \mathbf{M} \cdot \mathbf{v}' + \sigma, \tag{3.2}$$

where $\mathbf{v}' \equiv \mathbf{v} - \mathbf{v}_0(\mathbf{x})$ is a star's residual velocity, $\mathbf{M}(\mathbf{x})$ is a symmetric

31

matrix, and $\sigma(\mathbf{x})$ is a scalar function. That is, Eddington assumed f to be a function $f(Q)$. This hypothesis is called the ellipsoidal hypothesis because Q is constant on ellipsoids in velocity space. The principal axes of these velocity ellipsoids are specified by $\mathbf{M}$ and they are centered on the mean-streaming velocity at $\mathbf{x}$, namely $\mathbf{v}_0(\mathbf{x})$. The principal velocity dispersions are inversely proportional to the square roots of the eigenvalues of $\mathbf{M}$. The variation of the stellar density from point to point is largely controlled by σ. In [2] Chandrasekhar remarks that if several different models can be found that share a common gravitational potential, a more complex model that does not satisfy the ellipsoidal hypothesis can be formed by taking any weighted sum of the distribution functions of the basic models. He calls this idea the principle of superposition in stellar dynamics. Since the phenomenon of Oort's "high-velocity stars" informs us that the stellar velocity distribution at the Sun does not even approximately satisfy the ellipsoidal hypothesis, this principle is essential if the ellipsoidal hypothesis is to offer any hope of casting light on the dynamics of galaxies such as the Milky Way.

The ellipsoidal hypothesis requires that $Q + \sigma$ be a constant of stellar motion. So the question arises: which potentials admit several integrals that are quadratic in the velocities? This problem had been addressed by Whittaker (1936), who concluded that the potentials were those already identified by Stäckel (1883) as giving rise to separable Hamilton-Jacobi equations.

Strangely, in [1] Chandrasekhar does not proceed from Whittaker's work but formulates the problem anew and comes to the conclusion that if a model contains differential star-streaming, it must possess at least helical symmetry. Such a system can be spatially finite only if it is axially symmetric.

This result is surprising because Stäckel's potentials are not all axisymmetric. It is the more surprising in that Chandrasekhar argues that the ellipsoidal hypothesis allows a much larger range of solutions than Eddington had supposed possible, and Eddington had concluded that the allowed potentials were just Stäckel's potentials. Eddington did certainly err in assuming that the directions of the principal axes of the velocity ellipsoids at different points within the model would define a system of coordinate planes. As Chandrasekhar pointed out, the directions that a given principal axis takes at different points define a global vector field $\mathbf{n}(\mathbf{x})$, but an integrability condition must be satisfied if it is to be possible to solve the partial differential equation $\mathbf{n} \cdot \nabla \phi = 0$ for the function $\phi(\mathbf{x})$ which would define Eddington's coordinate surfaces through $\phi = $ constant.

Chandrasekhar sought solutions to his sets of equations by recasting them in all the usual coordinate systems. In this way he was able to characterize completely all planar and spherical solutions. For the planar case he derived an interesting model in which the mean azimuthal streaming velocity is given by

$$\overline{v}_\phi(R) = \frac{s}{\sqrt{s^2 + R^2}}, \tag{3.3}$$

and the principal velocity dispersions satisfy the Oort-Lindblad relation $\sigma_\phi^2/\sigma_R^2 = -B/(A - B)$, where A and B are the Oort constants.

Chandrasekhar's treatment of three-dimensional systems was less complete, but did turn up two remarkable homogeneous systems with spheroidal bounding surfaces within which the stellar streaming velocity has a component perpendicular to the equatorial plane. He speculated that the existence of one of these systems might be connected with the fact that the apparently flattest elliptical galaxies have axis ratio $a{:}c \simeq 3{:}1$.

In his second paper on the ellipsoidal hypothesis, Chandrasekhar allowed the quantities $\mathbf{v}_0$, $\mathbf{M}$ and σ to become functions of time as well as of position. This enabled him to demonstrate the instability of some of his earlier models, as well as to investigate models that display spiral structure.

His fundamental set of governing equations comprised a set of ten coupled partial differential equations in space that must be satisfied by the components of $\mathbf{M}$, six further equations connecting the time derivative of $\mathbf{M}$ to $\mathbf{v}_0(\mathbf{x}, t)$, and six integrability equations. He investigated solutions to this extremely complex system of equations under various assumptions regarding the form of the gravitational potential $\Phi(\mathbf{x}, t)$. For an axially symmetric system the latter has to be of the form

$$\Phi(r, \theta) = \Phi_0(t) - \frac{\ddot{s}}{2s} r^2 + \frac{1}{s^2} \Phi_1(r/s, \theta), \tag{3.4}$$

where Φ_0, Φ_1 and $s(t)$ are arbitrary functions. Chandrasekhar used this formalism to study the dynamics of homogeneous spheroidal systems in some detail. He found that such systems have both stable and unstable modes which move them between configurations that satisfy the ellipsoidal hypothesis. The unstable modes corresponded to uniform contraction or expansion of the system.

The last part of this massive paper discusses systems with isotropic distributions of residual velocities, so that the distribution function is of the form

$$f = f\left(s^2 |\mathbf{v} - \mathbf{v}_0|^2 + \sigma\right). \tag{3.5}$$

The structure of the model is determined by $s(t)$. If $\dot{s} > 0$, the spiral arms are leading and the system is expanding, and conversely if $\dot{s} < 0$. The azimuthal streaming velocity in these models is approximately of solid-body form.

3.3 The relaxation time

After his exhausting if not entirely exhaustive discussion of the ellipsoidal hypothesis, Chandrasekhar turned to the problem of stellar relaxation.

From the pioneering work of Jeans it was known that to lowest order it is possible to imagine that stars move in galaxies and globular clusters in the smoothed out potential that one obtains if one replaces the actual stellar distribution by the underlying probability density. In the next order of approximation one must allow for the deflections of stars from their zeroth-order orbits that are caused by the graininess of the actual potential. In [3] Chandrasekhar estimated the time t_r required for the cumulative effect of these deflections to become appreciable by calculating the time average of $\sum(\Delta E)^2$, where ΔE is the change in energy that a star suffers during a binary encounter with another star. This quantity had been earlier incorrectly evaluated by Eddington, K. Schwarzschild, and Rosseland. The intricacy of the calculation arises because distant, weak encounters cause the sum to diverge logarithmically with the impact parameter b_{max} of the most distant encounter considered. Chandrasekhar argued that b_{max} should be taken to be the mean distance between stars. Under this assumption he evaluated the relaxation time t_r in the Milky Way near the Sun and in a typical globular cluster, finding that t_r was $\sim 10^{14}$ yr in the former case and $\sim 10^{10}$ yr in the latter case.

In [4] Williamson and Chandrasekhar employed a new criterion to estimate the relaxation time t_r: they evaluated the time average of $\sum \sin^2 \phi$, where ϕ is the angle through which a star is deflected during a binary encounter. The resulting values of t_r were in good agreement with those obtained by consideration of $\sum(\Delta E)^2$ unless the test star was moving very much faster than the rms velocity of the field stars.

In [5] Chandrasekhar expresses profound dissatisfaction with the approach of [3] and [4]. This was brought to a head by an attempt to calculate the time average of $\sum \Delta E$. Since ΔE has a fluctuating sign, it is to be expected that the sum in question is the small difference of large terms. Mathematically, one has to evaluate an integral (over the parameters of binary orbits) that is not absolutely convergent. Worse still, to obtain a finite result for this integral, an upper limit b_{max} has to be set to the range of the impact parameter. Chandrasekhar feared that terms that depend strongly on b_{max} could not realistically be taken to cancel to the accuracy required if one is to obtain a physically plausible result. Since the problem was associated with the consideration of binary encounters at large impact parameters, he suspected that the underlying problem lay with the fundamental assumption that it was possible to treat the effects of graininess in the gravitational potential as the sum of a large number of entirely independent binary encounters.

Chandrasekhar's [6] contains his first attempt at a better theory of fluctuations in stellar systems. Holtsmark had already evaluated the probability distribution of the different values of the electric field at a given point in a plasma. Moreover, Smoluchowski had argued that the strength of a fluctuating random variable could be taken to relax exponentially back towards its equilibrium value after each upward fluctuation. The characteristic relaxation time would in general be a function $\tau(F)$ of the magnitude F of the original fluctuation. Chandrasekhar reasoned that he could use these results to calculate the relaxation time in a stellar system if he could calculate Smoluchowski's relaxation time $\tau(F)$.

Exasperatingly, this approach was still bedeviled by divergencies—this time associated with large field strengths due to close encounters. Chandrasekhar recognized that a satisfactory treatment of the problem of high field strengths lay beyond his present approach and simply imposed a lower cutoff on the interstellar distance. This cutoff was taken to be velocity dependent, being larger for lower relative velocities of stars.

Since Chandrasekhar could show that nearest neighbors dominate Holtsmark's probability distribution, he could estimate the lifetime of a given field strength as the time required by the perturbing neighbor to move significantly farther away or nearer.

Chandrasekhar's two papers [7] and [8] with von Neumann worked out in a rigorous way the ideas that lay behind [6]. In each paper the principal task was to evaluate the joint probability distribution

$$\frac{\mathrm{d}^6 P(\mathbf{F}, \dot{\mathbf{F}})}{\mathrm{d}^3\mathbf{F}\,\mathrm{d}^3\dot{\mathbf{F}}} \tag{3.6}$$

that a test star of mass M experiences a force $\mathbf{F}$ that changes at rate $\dot{\mathbf{F}}$. Paper [7] assumes an isotropic distribution of random velocities of perturbing stars relative to the test star. Since motion of the test star relative to the rest frame of a cluster as a whole will be reflected in an anisotropic distribution of velocities relative to that star, paper [8] evaluates $P(\mathbf{F}, \dot{\mathbf{F}})$ for the anisotropic case.

$\mathbf{F}$ is a function of the positions of all the stars: if $\mathbf{x}_i$ is the position of the ith field star, we have

$$\mathbf{F}(\mathbf{x}, \{\mathbf{x}_i\}) = \sum_i \frac{GMm_i(\mathbf{x} - \mathbf{x}_i)}{|\mathbf{x} - \mathbf{x}_i|^3} \equiv \sum_i \mathbf{F}_i. \tag{3.7}$$

Differentiating (3.7) with respect to t, one easily obtains a similar expression for $\dot{\mathbf{F}}(\mathbf{x}, \mathbf{v}, \{\mathbf{x}_i, \mathbf{v}_i\})$.

The probability P of finding our N field stars in any given region τ of $6N$-dimensional phase space is

$$P = \int_\tau \prod_{i=1}^N \left(\frac{1}{V}\,\mathrm{d}^3\mathbf{x}_i\,\mathrm{d}^3\mathbf{v}_i f(\mathbf{x}_i, \mathbf{v}_i)\right), \tag{3.8}$$

where f has been normalized so that

$$\int \mathrm{d}^3\mathbf{x}\,\mathrm{d}^3\mathbf{v} f = V. \tag{3.9}$$

Then from the standard properties of the Dirac δ-function it follows that

$$W(\mathbf{F}_0, \dot{\mathbf{F}}_0) \equiv \frac{\mathrm{d}^6 P(\mathbf{F}_0, \dot{\mathbf{F}}_0)}{\mathrm{d}^3 \mathbf{F}_0 \, \mathrm{d}^3 \dot{\mathbf{F}}_0}$$

$$= \int \prod_{i=1}^{N} \left(\frac{1}{V} \mathrm{d}^3 \mathbf{x}_i \, \mathrm{d}^3 \mathbf{v}_i \, f(\mathbf{x}_i, \mathbf{v}_i) \right)$$

$$\times \delta^3(\mathbf{F} - \mathbf{F}_0) \delta^3(\dot{\mathbf{F}} - \dot{\mathbf{F}}_0). \tag{3.10}$$

On replacing each δ-function by

$$\delta(x) = \int \frac{\mathrm{d}k}{2\pi} \, \mathrm{e}^{ikx} \tag{3.11}$$

we quickly find that the Fourier transform of W is

$$\widetilde{W}(\mathbf{k}, \mathbf{k}') = \left(\frac{1}{V} \int \mathrm{d}^3 \mathbf{x}_i \, \mathrm{d}^3 \mathbf{v}_i \, f(\mathbf{x}_i, \mathbf{v}_i) \mathrm{e}^{i(\mathbf{k} \cdot \mathbf{F}_i + \mathbf{k}' \cdot \dot{\mathbf{F}}_i)} \right)^N, \tag{3.12}$$

where f is the system's distribution function. Since we are interested in the limit $N \to \infty$, $N/V = \text{constant} \equiv n$, we exploit equation (3.9) to write

$$\widetilde{W}(\mathbf{k}, \mathbf{k}') = \left(1 - \frac{1}{V} C(\mathbf{k}, \mathbf{k}') \right)^{nV}$$

$$= \mathrm{e}^{-nC}, \tag{3.13}$$

where

$$C(\mathbf{k}, \mathbf{k}') \equiv \int \mathrm{d}^3 \mathbf{x}_i \, \mathrm{d}^3 \mathbf{v}_i \, f(\mathbf{x}_i, \mathbf{v}_i) \left(1 - \mathrm{e}^{i(\mathbf{k} \cdot \mathbf{F}_i + \mathbf{k}' \cdot \dot{\mathbf{F}}_i)} \right). \tag{3.14}$$

In a *tour de force*, Chandrasekhar and von Neumann did the six-dimensional integral in (3.14) by changing integration variables from $(\mathbf{x}_i, \mathbf{v}_i)$ to $(\mathbf{F}_i, \dot{\mathbf{F}}_i)$. They could then obtain W by inverse Fourier transformation and finally evaluate expectation values of interest.

In their first paper they tabulated the lifetimes of states with given values of F, where the lifetime was defined to be $T \equiv F/\sqrt{\langle \dot{F}^2 \rangle_F}$. In their second paper they derived the formula

$$\langle \dot{\mathbf{F}} \rangle_{\mathbf{F}, \mathbf{v}} = \frac{2}{3} \pi \, GmnB(F, m) \left(3 \frac{\mathbf{v} \cdot \mathbf{F}}{F^2} \mathbf{F} - \mathbf{v} \right). \tag{3.15}$$

Here m is the average mass of the field stars and B is a given positive function. Dotting through by $\mathbf{F}$ we find

$$\left\langle \frac{dF^2}{dt} \right\rangle_{\mathbf{F}, \mathbf{v}} = \frac{8}{3} \pi \, GmnB \, \mathbf{F} \cdot \mathbf{v}. \tag{3.16}$$

Hence **F** tends to increase in magnitude when the star is moving in the direction of **F**, and decrease in magnitude in the opposite case. Chandrasekhar and von Neumann argue that the existence of dynamical friction follows from this result. There would seem to be a more elementary and convincing physical interpretation of equation (3.16): **v** is in the same direction as **F** when a star is falling into a potential well. The potential of a point particle is such that then F is increasing. Thus equation (3.16) merely reflects the obvious kinematics of motion through a cloud of point masses.

In the first part of [9] Chandrasekhar showed that the existence of dynamical friction follows from the ability of a system to come to thermal equilibrium. Specifically, if a force of magnitude F acts for a time T, the star's velocity changes by an amount $|\delta\mathbf{v}| = FT$. After time t the star will have experienced $N = t/T$ such random velocity increments, and in velocity space will have diffused a distance of order $\Delta v = FT\sqrt{N} = F\sqrt{Tt}$. If the system is to come into thermal equilibrium, the diffusion to ever higher velocities has to be resisted by a frictional force. If, following the work of Ornstein, Uhlenbeck, and others on Brownian motion, one assumes that the resistive force is proportional to **v**, one concludes that its coefficient must be

$$\eta = \frac{\langle F^2\rangle T}{2\langle v^2\rangle}. \tag{3.17}$$

In the second part of [9] Chandrasekhar presented the now conventional derivation of dynamical friction: one averages, over all encounters, the component of the change in a star's velocity that is parallel to its peculiar velocity. Reassuringly, the resulting value of η satisfies equation (3.17).

In [10] Chandrasekhar estimates the rate at which stars escape from a star cluster. This he does by considering the diffusion of a star in velocity space. At $t = 0$ the probability density $f(v)$ of the star having speed v is $\delta(v - v_0)$, and at all times f is assumed to vanish at and above the escape velocity v_e. It is a simple matter to solve the relevant diffusion equation subject to these boundary conditions. Then the rate of escape at time t is just the integral of the probability current around the sphere of radius v_e.

In this way Chandrasekhar estimated the times required for galactic clusters to evaporate both when dynamical friction was included and when it was ignored. The inclusion of dynamical friction lengthened the time required for a cluster to evaporate by a factor of between 15 and 50 depending on the cluster's central concentration. In the specific case of the Pleiades cluster, the inclusion of dynamical friction extended the expected lifetime of the cluster from $\sim 3 \times 10^7$ yr to $\sim 5 \times 10^8$ yr.

In the final section of [10] Chandrasekhar derives the collisional term of the full Boltzmann equation for use in further studies. It is

$$\frac{\partial f}{\partial t} + \mathbf{v} \cdot \frac{\partial f}{\partial \mathbf{x}} - \frac{\partial \Phi}{\partial \mathbf{x}} \cdot \frac{\partial f}{\partial \mathbf{v}} = \frac{\partial}{\partial \mathbf{v}} \cdot \left(q \frac{\partial f}{\partial \mathbf{v}} + \eta f \mathbf{v} \right). \qquad (3.18)$$

Paper [11] upgrades the analysis of [10] by allowing for the velocity-dependence of the coefficients q and η in equation (3.18), which had been taken to be constants in [10]. This upgrade, which necessitated significant numerical work, increased the predicted lifetimes of clusters such as the Pleiades to values of order 3×10^9 yr.

In [12] Chandrasekhar applied the methodology developed in his papers with von Neumann to the calculation of the correlation function of the gravitational field. That is, he calculated several projections of the tensor-valued expectation value $\langle F_i^{(1)} F_j^{(2)} \rangle$, where $\mathbf{F}^{(i)}$ is the force at $\mathbf{x}^{(i)}$. The key quantity to be calculated is the characteristic function $C(\mathbf{k}, \mathbf{k}')$ that determines the probability density $W(\mathbf{F}^{(1)}, \mathbf{F}^{(2)})$ through equation (3.13). It is given by [cf. (3.14)]

$$C(\mathbf{k}, \mathbf{k}', \mathbf{\Delta}) = N \int d^3\mathbf{x}\, d^3\mathbf{v}\, f(\mathbf{x}, \mathbf{v}) \left(1 - e^{i[\mathbf{k} \cdot \mathbf{F}(\mathbf{x}) + \mathbf{k}' \cdot \mathbf{F}(\mathbf{x} + \mathbf{\Delta})]} \right). \qquad (3.19)$$

After evaluating the integral in (3.19), Chandrasekhar was able to show that the mean value of the field at $\mathbf{x} + \mathbf{\Delta}$ dotted with the unit vector in the direction of the field at $\mathbf{x}$ has a Taylor series of the form

$$\langle \hat{\mathbf{F}}^{(1)} \cdot \mathbf{F}^{(2)} \rangle = a_0 - a_1 \Delta + a_4 \Delta^4 + \cdots, \qquad (3.20)$$

where the a_i are positive quantities that he evaluated.

Paper [13] calculates the temporal autocorrelation $\langle F_i(t_1) F_j(t_2) \rangle$ of the force experienced by a test star at times t_1 and t_2. This problem is assumed to be identical with that solved in [12] upon substitution of $\mathbf{v}(t_2 - t_1)$ for $\mathbf{\Delta}$ in equation (3.19). That is, Chandrasekhar neglects the explicit dependence of $\mathbf{F}(\mathbf{x}, t)$ on time.

In [14] Chandrasekhar applies the equal-time autocorrelation of the field evaluated in [12] to the dissolution of wide binaries. His reasoning is as follows. He focuses on the relative acceleration of the two stars in the direction of the background force that acts on star 1. After a time t, this has changed the relative velocities of the two stars by an amount

$$\Delta v = [\mathbf{F}(\mathbf{x}^{(1)}) - \mathbf{F}(\mathbf{x}^{(2)})] \cdot \hat{\mathbf{F}}(\mathbf{x}^{(1)})t, \qquad (3.21)$$

where the hat indicates a unit vector. On taking the expectation of both sides of the equation, results proved in [12] allow one to show that the right-hand side equals $4\pi Gmnat$, where m and n are the mass and number density of the background stars and a is the separation of the binary components. Chandrasekhar argues that the binary will be dissolved once $4\pi Gmnat$ has become comparable to the orbital velocity of the binary, namely $[G(m_1 + m_2)/a]^{1/2}$. This analysis suggests that binaries with $a \gtrsim 2000\,\mathrm{AU}$ will be dissolved within the age of the Milky Way.

3.4   The book

The book *Principles of Stellar Dynamics* appeared in 1942, around the time of his first paper with von Neumann. After a brief survey of what was then known observationally about the Milky Way, clusters, and external galaxies, the book discusses the relaxation time along the lines of papers [3–5]. The next two chapters give highly condensed and significantly clarified versions of papers [1] and [2] before presenting Lindblad's ideas about spiral structure. The fifth and final chapter concerns star clusters. Global conservation theorems, including the virial theorem, are derived. The relaxation time and the rate of stellar evaporation are estimated. The effects of tides on the stability of a homogeneous ellipsoidal cluster that moves on a circular orbit around the galactic center are studied. Finally the equilibria of spherical clusters are considered from the point of view of the isothermal sphere.

3.5   The influence of Chandrasekhar's papers

3.5.1   The ellipsoidal hypothesis

The impact of Chandrasekhar's two papers on the ellipsoidal systems has necessarily been limited by the extraordinary length of these highly mathematical papers. The discussion is throughout of a mathematical rather than a physical nature.

The equilibrium of a stellar system is determined by the interplay of dynamics, which requires that the distribution function be a constant of stellar motion, and Poisson's equation. The remarkable thing about papers [1] and [2] is the infrequency with which Poisson's equation is mentioned. Implicitly it appears a few times as the underpinning for the quadratic nature of the potential of a homogeneous ellipsoid. But the analysis is dominated by the implications of f being a constant of motion.

The ellipsoidal hypothesis now seems a confusing amalgam of two logically distinct ideas. First, Jeans' theorem states that the distribution function depends on $(\mathbf{x}, \mathbf{v})$ only through constants of motion. Second, we ask, which potentials have isolating integrals that are up to quadratic functions of the velocities? Energy is always such an integral. Angular momentum is often another. For a century it has been known that the general steady-state potential with three global quadratic isolating integrals is a Stäckel potential. This knowledge did not make its full impact on stellar dynamics until 1985 when de Zeeuw (1985) showed that some remarkably realistic model galaxies have Stäckel potentials. Eddington's investigation of the ellipsoidal hypothesis brought him closer to the discovery of the astronomical importance of Stäckel potentials than did Chandrasekhar's.

Most of Chandrasekhar's work on the ellipsoidal hypothesis was concerned with time-dependent models. There are still extremely few results in this field. In fact, other than Chandrasekhar's, the only exact time-dependent models of which I am aware are those of Freeman (1966) and Sridhar (1989). Both of these were constructed by first identifying a nontrivial isolating integral of the potential. In many respects Chandrasekhar's work foreshadows these models, but does not seem to have influenced them directly.

3.5.2   Discreteness noise in the gravitational field

It is surely Chandrasekhar's papers on the effects of discreteness noise in the gravitational field that have been the most influential of his stellar-dynamical papers. The idea that stellar systems are nearly collisionless but should slowly relax as a result of discreteness effects had been understood for many years before Chandrasekhar entered the field. His contribution was to calculate the diffusion coefficients $\langle \Delta E \rangle$ and $\langle (\Delta E)^2 \rangle$ accurately, to state the collisional Boltzmann equation clearly, and, above all, to identify the action of dynamical friction.

With hindsight his path to dynamical friction seems tortuous. He first lost confidence in what we now regard as the standard way to calculate diffusion coefficients, including the coefficient of dynamical friction. Then with von Neumann he developed an approach to the study of discreteness noise that avoids the concept of a binary encounter, which Chandrasekhar had identified as the source of his unease with standard methodology. Dynamical friction makes its first appearance in stellar dynamics at the end of Chandrasekhar's second paper with von Neumann. This fact is remarkable, as I think it is clear that dynamical friction lies beyond the reach of the Chandrasekhar–von Neumann approach. Indeed, as Mulder (1983) has so elegantly described, the frictional drag that a body experiences as it moves through a stellar system arises from the attraction that the body experiences for the region of enhanced density that tails behind it like a wake behind a ship. By contrast, a basic assumption of the Chandrasekhar–von Neumann method is that stars are randomly distributed in real space. So there is no way that the frictional force could emerge from the calculations of paper [8].

The conclusion is inescapable that by the time [8] was being finished, Chandrasekhar had deduced the existence of dynamical friction by the arguments of [9] and looked retrospectively at his calculations with von Neumann for evidence of the phenomenon.

Two things are notable about Chandrasekhar's discussion of dynamical friction. First, there is no mention that it is caused by each particle being attracted backwards by its gravitational wake. Second, there is no mention of its implications for the dynamics of massive bodies. In the 1970s dynamical friction would be seen to play a large role in several interesting astronomical phenomena by opposing the motion of galaxies, star clusters, and interstellar gas clouds. But Chandrasekhar is so focused on the role that friction plays in the establishment of thermal equilibrium amongst bodies of comparable mass that he makes light of the remarkable fact that a body's deceleration is proportional to its mass.

Chandrasekhar's work on evaporation from star clusters remained fiducial until Hénon's work in the 1960s (e.g., Hénon 1960). The main respect in which it required refinement was its disregard of the increase in the periods of stars as their energies creep up towards the escape energy. Although Chandrasekhar was probably the first person to write down the collisional Boltzmann equation for a stellar system, his treatment of cluster dynamics was confined to velocity rather than phase space. Hence it could not take into account variations in the time stars spend in a cluster's collision-dominated core.

The effects of stochastic variations in astronomical gravitational fields remain an important area of research. The core of the field has grown directly out of the part of Chandrasekhar's work that rested on the contributions of Jeans and K. Schwarzschild. By contrast, the part of Chandrasekhar's work that was inspired by the work of Holtsmark has achieved little resonance. I think the reason is that while it provides a wealth of information regarding the spatial structure of the field, it provides at best very limited information about the temporal structure of the field. This is unfortunate because an elementary calculation shows that the rate of change of a star's energy is the integral along its path of $\partial\Phi/\partial t$. Thus classical relaxation is entirely determined by temporal variations in $\mathbf{F} = -\nabla\Phi$, and a technique such as that developed by Chandrasekhar and von Neumann, which provides at most the expectation of the first time derivative of the field, is not promising.

I am aware of one important problem in which much can be done from a knowledge of the spatial structure of Φ alone. This is the dynamics of centrifugally supported disks. Such disks tend to heat in the sense that stars diffuse away from nearly circular orbits, narrowly confined to the equatorial plane, onto more eccentric and/or highly inclined orbits. Such diffusion can take place at constant energy, so that the equation $\dot{E} = \partial\Phi/\partial t$ does not constrain the essential phenomenon in an important way. For further discussion of this problem, which throws up several evident connections with Chandrasekhar's work, see Binney and Lacey (1988).

Was Chandrasekhar right to profoundly suspect the use of binary encounters in the calculation of diffusion coefficients? It is now clear that his fears were exaggerated. Indeed, Theuns (1996) has shown that standard theory based on binary encounters allows an excellent quantitative understanding of stellar diffusion in N-body models. These simulations confirm Spitzer's conclusion (Spitzer 1987) that the upper limit b_{max} on the impact parameters that should be considered is not the mean interstellar distance, as Chandrasekhar supposed, but the distance within which the stellar density is comparable to its local value.

Current interest in the area of fluctuating gravitational fields centers on the degree to which collective oscillations lead to larger fluctuations than one would expect from two-particle noise alone—see, e.g., Weinberg (1994).

3.5.3 Wide binaries

Chandrasekhar was attracted to the question of the dissolution of wide binaries as an application of results on the fluctuation of gravitational fields that he had derived following his papers with von Neumann. Unfortunately, his calculation is not persuasive. It makes sense to write $\Delta v \simeq Ft$ only if the force $\mathbf{F}$ acts in the same direction throughout the interval $(0, t)$. Now Chandrasekhar equates Δv to the product of time and the difference in the external forces on the two stars dotted with the unit vector in the direction of the force that acts on one of them. Therefore he is implicitly assuming that this direction is constant. For consistency he should have assumed that the direction of the force on the other star is also constant. In this case, the directions of the two external forces would be fixed in the rotating frame of the binary. This is implausible. In reality there are two cases to consider. In the first case, the binary period is short compared to the characteristic time of fluctuations in the background field. Then each star will experience the average of the background field around its orbit, and the two stars will suffer very similar net accelerations. Mathematically, the orbital elements will be constant by virtue of their adiabatic invariance. In the opposite limit, the binary period is long compared to the characteristic time of fluctuations. This is the limit in which the orbit can be disrupted, but it is also the limit that Chandrasekhar could not address because for this problem he did not even have the expectation value of the field's first time derivative.

Subsequent work (e.g., Weinberg *et al.* 1985) has assumed that the dominant fluctuations arise from encounters between a binary and either a single star or a bound object such as a cluster or an interstellar cloud. With this assumption it is straightforward to evaluate the diffusion coefficients that are required to follow the evolution of any given initial population of binaries. This sort of analysis has the potential to place important constraints on the degree to which the Milky Way's dark matter is concentrated into massive objects. The main difficulty at the present time is the observational determination of the numbers of binaries in each range of semi-major axes.

3.6 Summary

Chandrasekhar was essentially an applied mathematician rather than a physicist, and one who was very "productive" in the modern administrator's use of the word. Both of these characteristics tended to diminish his impact on the field of stellar dynamics. His mathematical orientation ensured that his papers are heavy going for a theoretical astrophysicist and completely impenetrable to the average astronomer. After pages of detailed calculations of particular integrals, integrability conditions, and the roots of equations, one longs for relief in the form of the description of the physical picture which emerges from the mathematics. Too often one longs in vain.

The essential difficulty of much of his mathematics was surely compounded by the speed with which he published. Several notations are frequently used for essentially the same quantity at different points in a paper, and determining the meaning of an equation can involve an exhausting backward chase through long chains of definitions. The potential for condensation and clarification is made evident by a comparison of chapters 3 and 4 of *The Principles of Stellar Dynamics* with papers [1] and [2] from which they substantially derive.

Rightly, scientists are remembered for the best rather than the worst things in their *oeuvres*. So it is proper that we should remember Chandrasekhar for his contributions to the theory of cluster evolution: for understanding how equilibrium between stochastic excitation and dynamical friction is attained, and for estimating cluster relaxation and evaporation times. In other fields he achieved more, but most would be happy to have attained as much in any field as he did in this.

References

Chandrasekhar's papers

[1]   The dynamics of stellar systems, I–VIII, *Astrophys. J.*, **90**, 1 (1939).

[2]   The dynamics of stellar systems, IX–XIV, *Astrophys. J.*, **92**, 441 (1940).

[3]   The time of relaxation of stellar systems, I, *Astrophys. J.*, **93**, 285 (1941).

[4]   The time of relaxation of stellar systems, II (with J. E. Williamson), *Astrophys. J.*, **93**, 305 (1941).

[5]   The time of relaxation of stellar systems, III, *Astrophys. J.*, **93**, 323 (1941).

[6]   A statistical theory of stellar encounters, *Astrophys. J.*, **94**, 511 (1941).

[7]   The statistics of the gravitational field arising from a random distribution of stars, I. The speed of fluctuations (with J. von Neumann), *Astrophys. J.*, **95**, 489 (1942).

[8]   The statistics of the gravitational field arising from a random distribution of stars, II. The speed of fluctuation; dynamical friction; spatial correlations (with J. von Neumann), *Astrophys. J.*, **97**, 1 (1943).

[9]   Dynamical friction, I. General considerations: The coefficient of dynamical friction, *Astrophys. J.*, **97**, 255 (1943).

[10]   Dynamical friction, II. The rate of escape of stars from clusters and the evidence for the operation of dynamical friction, *Astrophys. J.*, **97**, 263 (1943).

[11]   Dynamical friction, III. A more exact theory of the rate of escape of stars from clusters, *Astrophys. J.*, **98**, 54 (1943).

[12]   The statistics of the gravitational field arising from a random distribution of stars, III. The correlations in the forces acting at two points separated by a finite distance, *Astrophys. J.*, **99**, 25 (1944).

[13]   The statistics of the gravitational field arising from a random distribution of stars, IV. The stochastic variation of the force acting on a star, *Astrophys. J.*, **99**, 47 (1944).

[14]   On the stability of binary systems, *Astrophys. J.*, **99**, 54 (1944).

Other works

Binney, J. J., Lacey, C. G. 1988, *Mon. Not. R. Astr. Soc.*, **230**, 597.

de Zeeuw, P. T. 1985, *Mon. Not. R. Astr. Soc.*, **216**, 273.

Eddington, A. S. 1915, *Mon. Not. R. Astr. Soc.*, **76**, 37.

Freeman, K. C. 1966, *Mon. Not. R. Astr. Soc.*, **134**, 1.

Hénon, M. 1960, *Ann. d'Ap.*, **23**, 467.

Mulder, W. A. 1983, *Astron. and Astrophys.*, **117**, 9.

Spitzer, L. 1987, *Dynamical Evolution of Globular Clusters* (Princeton; Princeton University Press).

Sridhar, S. 1989, *Mon. Not. R. Astr. Soc.*, **238**, 1159.

Stäckel, P. 1883, *Math. Ann.*, **42**, 537.

Theuns, T. 1996, *Mon. Not. R. Astr. Soc.*, **279**, 827.

Weinberg, M. D. 1994, *Astrophys. J.*, **421**, 481.

Weinberg, M. D., Shapiro, S. L., Wasserman, I. 1985, *Astrophys. J.*, **312**, 367.

Whittaker, E. T. 1936, *A Treatise on the Analytical Dynamics of Particles and Rigid Bodies* (Cambridge: Cambridge University Press).

4

Radiative Transfer

George B. Rybicki

4.1 Introduction

Chandrasekhar's work in radiative transfer was typical of his lifelong pattern of working intensively in some particular field of physics for a few years, publishing a treatise on his work, and then essentially leaving the field.

The main bulk of the work was published during the period 1944–48 in a series of twenty-four articles in the *Astrophysical Journal* under the general title "On the radiative equilibrium of a stellar atmosphere."[1] This series represented a remarkable period of creativity for Chandrasekhar, bringing a wealth of new results and insights, which redirected and invigorated research in radiative transfer for many years afterwards.

In accordance with his custom, Chandrasekhar capped his achievements with the publication in 1950 of his treatise *Radiative Transfer*[2] [37], after which he published only a half-dozen more papers in the field.[3] As was often the case with his treatises, RT is noteworthy for its clarity, elegance, and style. Even forty-five years later it remains an important reference and guide to many of the fundamental concepts and methods of radiative transfer. This is more than just a volume of collected papers: his previous work is reorganized, with new insights added, and often with a simpler, more compact formulation.

Chandrasekhar never intended RT to be considered the last word on the subject, or his departure from the field to indicate that there was no more to be done. While RT had brought radiative transfer theory to a new high level, it left many clear challenges for the succeeding generation of workers in the field. As to his departure, his time to move on simply had come.

---

[1] Specified here by the symbol RE followed by a roman numeral, e.g., reference [23] is denoted RE-XIII. Many of these papers are reprinted in the collection [49]; this is indicated in the reference list by the notation, e.g., [SP2-41], for paper 41.

[2] Hereinafter RT.

[3] [38], [39], [40], [41], [43], and [50].

The purpose of this review is to give an overall feeling for the scope of Chandrasekhar's work and some appreciation of its significance. It is directed mainly at those who have some familiarity with radiative transfer theory, but who are not necessarily experts in it.

Because of the dominance of his treatise RT, the original papers of the "... Radiative equilibrium ..." series are seldom referenced today; RT has become the reference of choice for Chandrasekhar's work. Accordingly, the original papers will be cited here only when this seems necessary for understanding the historical and motivational aspects of Chandrasekhar's work. For convenience and consistency, the notation of RT will be adopted throughout, even if this was not the notation of the original papers. Some discussion of alternate notations will be given below, where appropriate.

In section 4.2 a brief presentation of some relevant radiative transfer theory is given. This is to provide a convenient reference point for the subsequent discussion and is not intended as a full introduction to the subject. Those desiring fuller explanations or more complete derivations should consult RT.

The major part of this review in sections 4.3–4.4 will focus on what are generally regarded as Chandrasekhar's three areas of greatest contribution, his work on: the discrete ordinate method, the invariance principles, and polarization. This division into separate areas is for convenience and should not be taken too literally, since the developments often proceeded in parallel, and results in one area often influenced another. For example, even after having developed his invariance techniques rather fully, he still needed considerable guidance in choosing appropriate forms for the solutions, and for this he often depended on results he had previously obtained using the discrete ordinate method. In section 4.6 a brief discussion is given on some of the influences of Chandrasekhar's work on the succeeding development of the field.

Chandrasekhar's work on radiative transfer theory represents a formidable achievement, one in which he took particular pride. He spoke fondly of this period:

> My research on radiative transfer gave me the most satisfaction.
> I worked on it for five years, and the subject, I felt, developed
> on its own initiative and momentum. Problems arose one by
> one, each more complex and difficult than the previous one, and
> they were solved. The whole subject attained an elegance and
> a beauty which I do not find to the same degree in any of my
> other work.[4]

It is hoped that this review can convey some of the "elegance and beauty" Chandraskhar brought to the field of radiative transfer.

---

[4] Quoted by Wali [51, p. 190].

4.2 Background

Problems in radiative transfer come in a bewildering number of forms, reflecting the various physical processes that may operate in different circumstances. In some cases, where the source function can be specified *a priori*, the problem is almost trivially solved using the so-called *formal solution* (see, e.g., RT, section 7). Apart from such special simple cases, radiative transfer problems typically involve *scattering*, which implies a source function that itself depends on the radiation field. This leads mathematically to an *integro-differential* equation of transfer. For these cases of scattering, the formal solution does not provide an explicit solution, although it may be used to reformulate the problem as an integral equation.

The great complexity of radiative transfer problems led early workers to concentrate on simple prototypical problems, for which some analytic progress might be made. Perhaps the simplest nontrivial geometry is *plane parallel*, in which all physical variables depend on only one Cartesian coordinate, say z (0 at the surface and measured inwards); this dependence is often specified instead by the normal optical depth τ. The direction of the ray can then be specified by the two spherical polar variables, θ, the angle of the ray with respect to the outward normal, and the corresponding azimuthal angle φ. It turns out to be very convenient to use the variable $\mu = \cos\theta$, $-1 \leq \mu \leq 1$, instead of θ itself. The monochromatic specific intensity at frequency ν then depends on just three variables: $I_\nu(\tau, \mu, \varphi)$.

Another common simplification is that the frequency of the radiation is unchanged upon scattering, the case of *elastic* or *monochromatic scattering*. In this case the frequency can be treated as a parameter, and the specific intensity can be written $I(\tau, \mu, \varphi)$.

Much of the character of general radiative transfer problems already appears in what is perhaps the simplest example, the case of unpolarized radiation with isotropic scattering in plane-parallel geometry with axial symmetry. In this case the radiative transfer equation is

$$\mu\frac{\partial I(\tau, \mu)}{\partial\tau} = I(\tau, \mu) - \frac{\varpi_0}{2}\int_{-1}^{1} I(\tau, \mu')\,d\mu', \tag{4.1}$$

where ϖ_0 has the important meaning of the fraction of radiation that is scattered (as opposed to absorbed), and is called the *single scattering albedo*. Accordingly, the case $\varpi_0 = 1$ refers to *conservative scattering*. Because of the axial symmetry, the azimuthal angle φ does not appear here.

For the more general case of anisotropic scattering, the transfer equation takes the form

$$\mu\frac{\partial I(\tau, \mu, \varphi)}{\partial\tau} = I(\tau, \mu, \varphi) - \frac{1}{4\pi}\int_{-1}^{+1}\int_{0}^{2\pi} p(\mu, \varphi; \mu', \varphi')I(\tau, \mu', \varphi')\,d\mu'd\varphi'$$

$$\tag{4.2}$$

(RT, p. 13, eq. 71). Here the *phase function* $p(\mu, \varphi; \mu', \varphi')$ describes the scattering from direction (μ', φ') into direction (μ, φ). Much of Chandrasekhar's work was concerned with the simple transfer equation (4.2). This equation shows the typical influence of scattering, in that the specific intensity in one direction depends on the specific intensity in all others. It is this implicit coupling that gives transfer problems their special difficulty.

For randomly oriented (nonaligned) scatterers, the phase function is a function of the cosine of the scattering angle Θ alone, that is,

$$\cos\Theta = \mu\mu' + (1 - \mu^2)^{1/2}(1 - \mu'^2)^{1/2}\cos(\varphi - \varphi'). \tag{4.3}$$

Then the general phase function can be conveniently expressed as a series in Legendre polynomials

$$p(\cos\Theta) = \sum_{l=0}^{\infty} \varpi_l P_l(\cos\Theta) \tag{4.4}$$

(RT, p. 7, eq. 33). The coefficient ϖ_0 again has the meaning of the single scattering albedo, and $\varpi_0 = 1$ refers to conservative scattering. A simplification in all of Chandrasekhar's work is that the coefficients ϖ_l are all independent of depth, the *homogeneous* case.

Among the cases treated by Chandrasekhar were the *isotropic scattering phase function*

$$p(\cos\Theta) = \varpi_0, \tag{4.5}$$

the *linear phase function*

$$p(\cos\Theta) = \varpi_0(1 + x\cos\Theta), \tag{4.6}$$

and *Rayleigh's phase function*

$$p(\cos\Theta) = \frac{3}{4}(1 + \cos^2\Theta). \tag{4.7}$$

It is necessary to distinguish between problems involving scattering of unpolarized radiation in accordance with the Rayleigh's phase function (4.7) and full *Rayleigh scattering* which includes the treatment of polarization (see section 4.5).

The geometries considered by Chandrasekhar were either *semi-infinite*, extending in optical depth from $\tau = 0$ to $\tau = \infty$, or *finite*, extending from $\tau = 0$ to $\tau = \tau_1$.

The various classes of transfer problems are now defined by their particular boundary conditions on the intensities. One important example is the *radiative equilibrium problem* or *Milne problem*, which is defined as a semi-infinite, conservative atmosphere with no incident intensity at $\tau = 0$, and with a condition at infinity that the intensities should not grow exponentially for any $\epsilon > 0$:

$$
\begin{aligned}
I(0, \mu, \varphi) &= 0, & -1 \leq \mu \leq 0 \\
I(\tau, \mu, \varphi)e^{\epsilon\tau} &\to 0, & \tau \to \infty.
\end{aligned} \tag{4.8}
$$

Because the problem is homogeneous, one can also specify the (constant) net flux F carried in the medium. The radiative equilibrium problem provides one of the simplest models for a stellar atmosphere.

Another important class of problems involves the *diffuse transmission and reflection* of radiation for a finite medium, or just the diffuse reflection for the semi-infinite medium. With radiation incident on $\tau = 0$ in direction $(-\mu_0, \varphi_0)$, and with net flux πF normal to the direction of the beam, the problem is to find the emergent diffuse intensities at the boundaries.

The term *diffuse* in this context refers to the separation of the radiation field into the unscattered part traveling in direction $(-\mu_0, \varphi_0)$ and the remaining so-called diffuse part, which has scattered at least once. The diffuse radiation is much smoother in its angular dependence, and so is more suitable to be treated by the discrete ordinate method, for example. It is easily shown that the diffuse radiation field satisfies a modified integro-differential equation:

$$
\begin{aligned}
\mu\frac{\partial I(\tau, \mu, \varphi)}{\partial \tau} &= I(\tau, \mu, \varphi) - \frac{1}{4\pi} \int_{-1}^{+1} \int_0^{2\pi} p(\mu, \varphi; \mu', \varphi')I(\tau, \mu', \varphi')\, d\mu' d\varphi' \\
&\quad - \frac{1}{4}Fe^{-\tau/\mu_0}p(\mu, \varphi; -\mu_0, \varphi_0),
\end{aligned} \tag{4.9}
$$

(RT, p. 22, eq. 126). The incident radiation is now fully accounted for in the inhomogeneous term, and the boundary conditions are for zero incident diffuse radiation at the boundaries.

The *diffuse scattering and transmission functions*, S and T, are now defined in terms of the emergent intensities at each boundary:

$$
\begin{aligned}
I(0, \mu, \varphi) &= \frac{F}{4\pi}S(\tau_1; \mu, \varphi; \mu_0, \varphi_0), \\
I(\tau_1, -\mu, \varphi) &= \frac{F}{4\pi}T(\tau_1; \mu, \varphi; \mu_0, \varphi_0).
\end{aligned} \tag{4.10}
$$

For the semi-infinite case no transmission function is defined. If the incident radiation has axial symmetry, only simplified versions of the scattering and transmission functions are required. These simplified functions do not depend on the azimuthal variables, which are then denoted[5] $S(\mu, \mu_0)$ and $T(\mu, \mu_0)$.

---

[5] The argument τ_1 is often omitted if it is clear from context.

Using these scattering and transmission functions, along with the linear superposition principle, it is possible to express the reflected and transmitted intensities corresponding to any arbitrary incident radiation. As will be seen below, the scattering and transmission functions are not merely of interest in special circumstances, but define the fundamental structure of the radiative transfer properties of the medium, and play a crucial role in the invariance principles.

4.3 The discrete ordinate method

An integro-differential equation of transfer with two-point boundary conditions (e.g., eq. [4.1] or [4.2]) presents a difficult mathematical problem. By 1944, fully analytic solutions had been presented for only a few problems (e.g., the solution to the semi-infinite Milne problem by Wiener and Hopf [3]), but these methods did not extend to all of the problems that were of interest to Chandrasekhar. This led him to adopt a scheme, introduced earlier by Wick [6], that reduced an integro-differential equation to an approximate, finite set of ordinary differential equations by the introduction of a quadrature scheme into the integral term. Because of Chandrasekhar's subsequent extensive development of this method, it is now often known as the *Wick-Chandrasekhar discrete ordinate method* or simply the *discrete ordinate method*.

The quadrature formula used in the Wick-Chandrasekhar method approximates the integral of an arbitrary function $f(\mu)$ from -1 to $+1$ by a sum over discrete values of the function:

$$\int_{-1}^{1} f(\mu)\, d\mu = \sum_{j} a_j f(\mu_j). \qquad (4.11)$$

The quadrature constants a_j and μ_j can be chosen in a variety of ways. Wick [6] suggested that the best choices for the constants are those of the Gaussian quadrature formula, for which the formula (4.11) is *exact* for any polynomial in μ of degree less than $(2n - 1)$. Because of the symmetry between the positive and negative values of μ, by restricting the order of quadrature to be even, say $2n$, these constants can be numbered in such a way that j takes the values $\pm 1, \pm 2, \ldots, \pm n$ where $a_{-j} = a_j$ and $\mu_{-j} = \mu_j$.

Applying the above quadrature formula to the integral in equation (4.1), and setting $\mu = \mu_i$, $i = \pm 1, \pm 2, \ldots, \pm n$, one obtains the equations of the discrete ordinate method:

$$\mu_i \frac{dI_i}{d\tau} = I_i - \frac{\varpi_0}{2} \sum_{j} a_j I_j, \qquad (4.12)$$

where $I_j = I(\tau, \mu_j)$. These represent a system of $2n$ coupled, ordinary differential equations for the components of specific intensity at the discrete angles μ_i. Solution of these equations is expected to yield an approximation to the true solution of the full integro-differential equation (4.1).

The discrete ordinate method gave highly accurate solutions in a surprisingly compact form. Consider, for example, the radiative equilibrium problem (Milne problem) for a constant flux F, where $\varpi_0 = 1$. Of particular interest is the angular distribution of the emergent intensity for this problem, which can be expressed as

$$I(0, \mu) = \frac{\sqrt{3}}{4} FH(\mu), \tag{4.13}$$

which defines $H(\mu)$, the H-function.

By use of the discrete ordinate equations of order n, Wick showed that the H-function could be expressed simply. As a preliminary step, one finds the n positive roots[6] k_α for k of the *characteristic equation*

$$1 = \frac{\varpi_0}{2} \sum_{j=1}^{n} \frac{a_j}{1 - \mu_j^2 k^2}. \tag{4.14}$$

In principle, this is equivalent to solving the n-order polynomial equation in k^2 that results upon multiplying both sides by the product of denominators. The H-function is then given by the simple formula

$$H(\mu) = \frac{\displaystyle\prod_{i=1}^{n} (1 + \mu/\mu_i)}{\displaystyle\prod_{\alpha=1}^{n} (1 + \mu k_\alpha)}. \tag{4.15}$$

It is interesting to compare this approximate expression for $H(\mu)$ to the earlier exact analytic formula for the Milne problem, due to Hopf [4, p. 105]:

$$H(\mu) = (1 + \mu) \exp\left\{ -\frac{\mu}{\pi} \int_0^{\pi/2} \frac{\log[(1 - \phi \cot \phi)/\sin^2 \phi]}{\cos^2 \phi + \mu^2 \sin^2 \phi} \, d\phi \right\}. \tag{4.16}$$

While this exact formula is beautifully elegant in its own right, it does require a separate numerical quadrature to determine the value of $H(\mu)$ for each value of μ. By contrast, the discrete ordinate method requires the solution to the characteristic equation (4.14), but once this is done the formula (4.15) is a simple closed expression for $H(\mu)$ for any value of μ.

Chandrasekhar's numerical comparison of low order results with the exact analytic result (4.16) convinced him that the approximate results of the discrete ordinate method would converge to the exact results in the

---

[6] For $\varpi_0 = 1$ there are only $n - 1$ such roots.

limit $n \to \infty$. Only much later was this convergence proved mathematically [44].

Chandrasekhar also applied the method of discrete ordinates to the problem of diffuse reflection, in which radiation is incident on the medium at angle μ_0, and one is required to find the radiation emergent at angle μ. This relationship is given in terms of a scattering function $S(\mu, \mu_0)$. It had previously been shown by Hopf [4, eq. 191] that the scattering function is related to the above H-function (4.16) by means of the relation

$$\left(\frac{1}{\mu} + \frac{1}{\mu_0}\right) S(\mu, \mu_0) = \varpi_0 H(\mu) H(\mu_0). \qquad (4.17)$$

Thus, the function $S(\mu, \mu_0)$ of *two* variables can be simply expressed in terms of a function of a *single* variable, the same H-function that appears in the solution for the radiative equilibrium problem.

When Chandrasekhar applied the discrete ordinate method to the semi-infinite diffuse reflection problem (cf. RT, section 26), he found a result of the same form as equation (4.17), where the H-function was precisely the same as that for the discrete ordinate solution (4.15) to the radiative equilibrium problem.

For the case of a finite medium, besides the diffuse scattering function $S(\mu, \mu_0)$ there is also a diffuse transmission function $T(\mu, \mu_0)$ to be determined. These functions satisfy the extended relations, given by Ambartsumian [5],

$$\left(\frac{1}{\mu} + \frac{1}{\mu_0}\right) S(\mu, \mu_0) = \varpi_0 \left[X(\mu)X(\mu_0) - Y(\mu)Y(\mu_0)\right],$$
$$\left(\frac{1}{\mu} - \frac{1}{\mu_0}\right) T(\mu, \mu_0) = \varpi_0 \left[X(\mu)Y(\mu_0) - Y(\mu)X(\mu_0)\right], \qquad (4.18)$$

where the functions[7] X and Y were solutions to certain functional equations. After seeing these forms in Ambartsumian's paper, Chandrasekhar (RE-XXI [31]) was able to put the discrete ordinate solution for this problem into the same form, where the X- and Y-functions were also expressible in closed form in terms of the roots of the characteristic function. This solution required almost fifteen pages of algebra in RE-XXI (shortened somewhat in RT, section 59). Anyone who has tried to reproduce this result, starting from scratch and properly arranging the vast arrays of equations into intelligible form, will gain tremendous respect for Chandrasekhar's feat of complex algebraic manipulation.

It may seem strange, even paradoxical, that Chandrasekhar's work in radiative transfer, rightly regarded as a *tour de force* in mathematical

---

[7]Ambartsumian used the notations ϕ and ψ for these functions.

physics, should have depended so strongly and intimately on a numerical approximation scheme such as the discrete ordinate method. But one should remember that the problems confronted by mathematical physicists, whatever the field, are usually highly idealized. The transfer problems Chandrasekhar considered had already been simplified by making a number of physical assumptions and approximations: e.g., plane-parallel geometry, coherent scattering, and single-scattering albedo independent of depth. In a sense, choosing discrete angles is just one more simplifying approximation, on a par with the others made. Then the crucial questions to ask are whether these simplified equations are of practical use, can increase our mathematical or physical understanding, or can satisfy some criterion of mathematical beauty. I believe for the discrete ordinate method the answer is "yes" to each of these questions.

As to the practicality of the method, remember that Chandrasekhar was not acting solely as a mathematical physicist, but as an astrophysicist attempting to find answers to practical problems in stellar atmospheres and planetary atmospheres. Much of his effort was devoted to the construction of detailed tables of functions to be used for solving real problems. Martin Schwarzschild said of him:[8]

> Chandra had no snobbishness in regard to his mathematical work. He did not shy away from numerical, computational solutions. He mixed rigorous analysis with numerical calculations, as the problem required.

The discrete ordinate method gave him highly accurate solutions in a completely straightforward way.

As to the method's relation to mathematical and physical understanding, Chandrasekhar was obviously delighted when he continually found many results of this method that were perfectly consistent with exact requirements of the theory. Many known, analytically exact results are obeyed precisely to all orders of approximation in the discrete ordinate method: for example, the Hopf-Bronstein relation $J(0) = \sqrt{3}F/4$ in the Milne problem, the structure of the diffuse scattering and transmission functions as given in equations (4.17) and (4.18), and their reciprocity relations. Often he was able to determine the form of the exact solutions only after he had solved the discrete ordinate equations first. These circumstances convinced Chandrasekhar that the discrete ordinate method was more than just a convenient numerical method; it also preserved essential mathematical and physical characteristics of the problem being investigated.

Mathematical beauty is a highy personal matter, and many would not apply such a term to the discrete ordinate method. However, in its defense

---

[8]Quoted by Wali [51, p. 188].

one notes that many of the basic results of the method are expressible in forms that are surprisingly compact and elegant, much more so than one might have imagined when first tackling a complicated set of coupled differential equations. It is clear that a great proportion of Chandrasekhar's day-to-day work in radiative transfer must have been involved with manipulations of quite complex sets of discrete ordinate equations. One can only assume that he found this perfectly consistent with his view of the field as one of elegance and beauty.

4.4   Invariance principles

An important component of Chandrasekhar's work in radiative transfer was devoted to the development of the *invariance principles*. These principles were first introduced by Ambartsumian in 1943 [5] and 1944 [7], but because of the difficulties in communications during World War II, Chandrasekhar did not immediately learn of this work. In [50] he relates how he became aware of [5] during the summer of 1945, and became aware only "very much later" of [7].

After seeing Ambartsumian's work [5] in 1945, Chandrasekhar quickly realized the importance of the invariance principles. Besides their intrinsic mathematical elegance, the invariance principles greatly simplified radiative transfer problems by showing from the outset the underlying structure of the solutions, e.g., the forms of the scattering and transmission functions in equations (4.17) and (4.18). Subsequently, Chandrasekhar began a substantial redirection of his efforts in radiative transfer. He spent much of the next years in presenting expanded derivations of the invariance principles and generalizing them to more complex cases, ultimately to include polarization.

In the context of plane-parallel radiative transfer problems, the invariance principles result from consideration of what happens upon addition (or subtraction) of layers of material to (or from) the surfaces of the medium. In the simplest case, the medium has constant properties, such as phase function and single scattering albedo, and the added layer shares these same properties.

Ambartsumian [5] considered the problem of diffuse reflection from an isotropically scattering semi-infinite medium. He observed the obvious physical fact that the scattering function would be unchanged if a small layer of material were added at the surface, since the resulting medium would still be semi-infinite. By simply tracking through the changes in intensities due to the added layer, and setting the net change to zero, he found that the scattering function became immediately expressible in the form given in equation (4.17). Thus the remarkable factorization of the scattering function in the semi-infinite case is a direct result of an invariance principle.

In the same paper [5], Ambartsumian also considered the appropriate generalization to the invariance principle for a finite medium. In this case, the reflection and transmission functions are unchanged by adding an infinitesimal layer to one surface while simultaneously subtracting a layer of the same thickness from the other surface. Without giving details, Ambartsumian stated that, as a result of this invariance principle, the reflection and transmission functions for the finite case were expressible in the simple forms given in equation (4.18). Using the Ambartsumian invariance results for guidance, Chandrasekhar was able in RE-XXI [31] to solve the discrete ordinate solution for isotropic scattering in a finite medium. He then extended the theory to include various cases of anisotropic scattering (RT, sections 64–65).

At this time Chandrasekhar began his own investigations into the invariance principles, which extended the groundbreaking work of Ambartsumian. The foundations of his work were four invariance principles, which were presented first in RE-XVII [27], and later, in their most elegant form, in RT (pp. 161–166, section 50). Because of their importance and beauty, it is worthwhile here to write Chandrasekhar's four principles out in full. They are expressed first in words, and then mathematically:

I. *The intensity, $I(\tau, +\mu, \varphi)$, in the outward direction at any level τ results from the reflection of the reduced incident flux $\pi F e^{-\tau/\mu_0}$ and the diffuse radiation $I(\tau, -\mu', \varphi')$ $(0 < \mu' \leq 1)$ incident on the surface τ, by the atmosphere of optical thickness $(\tau_1 - \tau)$, below τ.*

$$I(\tau, +\mu, \varphi) =$$
$$\frac{F}{4\pi} e^{-\tau/\mu_0} S(\tau_1 - \tau; \mu, \varphi; \mu_0, \varphi_0)$$
$$+ \frac{1}{4\pi\mu} \int_0^1 \int_0^{2\pi} S(\tau_1 - \tau; \mu, \varphi; \mu', \varphi') I(\tau, -\mu', \varphi') \, d\mu' d\varphi'. \quad (4.19)$$

II. *The intensity, $I(\tau, -\mu, \varphi)$, in the inward direction at any level τ results from the transmission of the incident flux by the atmosphere of optical thickness τ, above the surface τ, and the reflection by this same surface of the diffuse radiation $I(\tau, +\mu', \varphi')$ $(0 \leq \mu' \leq 1)$ incident on it from below.*

$$I(\tau, -\mu, \varphi) =$$
$$\frac{F}{4\pi} T(\tau; \mu, \varphi; \mu_0, \varphi_0)$$
$$+ \frac{1}{4\pi\mu} \int_0^1 \int_0^{2\pi} S(\tau; \mu, \varphi; \mu', \varphi') I(\tau, +\mu', \varphi') \, d\mu' d\varphi'. \quad (4.20)$$

III. *The diffuse reflection of the incident light by the entire atmosphere is equivalent to the reflection by the part of the atmosphere of optical thickness τ, above the level τ, and the transmission by this same atmosphere of the* *diffuse radiation* $I(\tau, +\mu', \varphi')$ $(0 \le \mu' \le 1)$ *incident on the surface* τ *from below.*

$$\frac{F}{4\pi} S(\tau_1; \mu, \varphi; \mu_0, \varphi_0) =$$

$$\frac{F}{4\pi} S(\tau_1 - \tau; \mu, \varphi; \mu_0, \varphi_0) + e^{-\tau/\mu} I(\tau, +\mu, \varphi)$$

$$+ \frac{1}{4\pi\mu} \int_0^1 \int_0^{2\pi} T(\tau; \mu, \varphi; \mu', \varphi') I(\tau, +\mu', \varphi') \, d\mu' d\varphi'. \quad (4.21)$$

IV. *The diffuse transmission of the incident light by the entire atmosphere is equivalent to the transmission of the reduced incident flux* $\pi F e^{-\tau/\mu_0}$ *and the diffuse radiation* $I(\tau, -\mu', \varphi')$ $(0 < \mu' \le 1)$ *incident on the surface* τ *by the atmosphere of optical thickness* $(\tau_1 - \tau)$ *below* τ.

$$\frac{F}{4\pi} T(\tau_1; \mu, \varphi; \mu_0, \varphi_0) =$$

$$\frac{F}{4\pi} e^{-\tau/\mu_0} T(\tau_1 - \tau; \mu, \varphi; \mu_0, \varphi_0) + e^{-(\tau_1-\tau)/\mu} I(\tau, -\mu, \varphi)$$

$$+ \frac{1}{4\pi\mu} \int_0^1 \int_0^{2\pi} T(\tau_1 - \tau; \mu, \varphi; \mu', \varphi') I(\tau, -\mu', \varphi') \, d\mu' d\varphi'. \quad (4.22)$$

These invariance principles were introduced by Chandrasekhar purely on intuitive and physical grounds; he made no attempt to prove them, starting, for example, with the transfer equation.

By differentiating equations (4.19)–(4.22) with respect to τ and passing to an appropriate limit, either $\tau = 0$ or $\tau = \tau_1$, and with some further manipulations, Chandrasekhar found a set of four integral relations involving the scattering and transmission functions and their derivatives with respect to τ_1 (RT, section 51). By eliminating the derivatives between two pairs of these equations, the two invariance principles of Ambartsumian were recovered. But this left two relationships involving the derivatives, so Chandrasekhar's procedure had in essence doubled the number of known invariance principles.

For example, in the case of isotropic scattering in an axially symmetric, finite atmosphere, two of these four relationships are just Ambartsumian's results previously given in equation (4.18). In addition, there are two new results:

$$\frac{\partial S(\tau_1; \mu, \varphi)}{\partial \tau_1} = \varpi_0 Y(\mu) Y(\mu_0),$$

$$\left(\frac{1}{\mu_0} - \frac{1}{\mu} \right) \frac{\partial T(\tau_1; \mu, \varphi)}{\partial \tau_1} = \varpi_0 \left[\frac{1}{\mu_0} X(\mu) Y(\mu_0) - Y(\mu) X(\mu_0) \right]. \quad (4.23)$$

It became apparent to Chandrasekhar that the invariance principles were a powerful tool for attacking complex problems, in part, simply by showing the structure of the solutions. It also became clear that X- and Y-functions were important, fundamental quantities, and he spent much effort in developing a theory for their properties. This theory is concerned with more general functions than the particular ones appropriate to the isotropic scattering problem. What Chandrasekhar studied were the class of functions that satisfied the pair of equations

$$X(\mu) = 1 + \mu \int_0^1 \frac{\Psi(\mu')}{\mu + \mu'} \left[X(\mu)X(\mu') - Y(\mu)Y(\mu') \right],$$

$$Y(\mu) = e^{-\tau_1/\mu} + \mu \int_0^1 \frac{\Psi(\mu')}{\mu + \mu'} \left[Y(\mu)X(\mu') - X(\mu)Y(\mu') \right]. \quad (4.24)$$

The *characteristic function* $\Psi(\mu)$ is an even function of μ (Chandrasekhar took this to be a polynomial, although much of his development is not dependent on that assumption). For the particular case $\Psi(\mu) = 1/2$, the X- and Y-functions defined by these equations are just the ones previously defined for the isotropic scattering problem.

The theory of the X- and Y-functions is very elegantly presented in RT, chapter 8. A great number of results are given there concerning these functions and their moments, which play a crucial role in Chandrasekhar's subsequent work. One problem that Chandrasekhar left unresolved in his theory of the X- and Y-functions was their nonuniqueness in the conservative case (RT, section 58). This was not settled until many years later by Mullikin [46].

The general invariance equations imply that many radiative transfer problems of great complexity, such as those involving anisotropic scattering or polarization, can be drastically simplified. Although the algebra was very difficult, he showed how the scattering and transmission functions could be reduced to expressions involving a number of trigonometric functions of $(\varphi - \varphi_0)$, plus a finite set of pairs of X- and Y-functions, each one with a different characteristic function $\Psi(\mu)$. Thus, once the X- and Y-functions are determined, these radiative transfer problems are *completely* solved.

The magnitude of this reduction is impressive. For fixed τ_1, it is equivalent to reducing the problem of solving for functions S and T of *three* variables, μ, μ_0, and $(\varphi - \varphi_0)$, to one of solving for functions of only *one* variable. In the case of Rayleigh scattering, Chandrasekhar achieved this reduction for each element of the scattering and transmission matrices for the Stokes parameters. Since the discrete ordinate method yields explicit formulas for the X- and Y-functions, Chandrasekhar could legitimately claim to have given complete and exact solutions to these very difficult problems.

Even with the insights provided by the invariance principles, these problems were still remarkably difficult. In order to solve them, Chandrasekhar needed to pick the correct form of the solution, and this he did by a combination of intuitive guessing and comparison with similar solutions found with the discrete ordinate method. Looking at one of his papers such as the sixty-four-page RE-XXI [31], one will also conclude that he possessed a heroic capacity for doing algebra.

4.5   Polarization

One of the motivations for Chandrasekhar's investigations into radiative transfer was the problem of the polarization of the sunlit sky. This problem had been treated by Lord Rayleigh, who first derived and explained the basic molecular scattering mechanisms of the atmosphere. However, Rayleigh's detailed predictions of sky polarization were based on the approximation of a single scattering of radiation and, while these matched the observations in very broad outline, substantial unexplained deviations still remained. For example, Rayleigh's simple theory predicted a nonvanishing amount of polarization in all directions (except directly towards or away from the Sun). However, it was known that, depending on the sun's location, there exist two (sometimes three) neutral points of zero polarization on the Sun's meridian circle, called the *Babinet*, *Brewster*, and *Arago* points. It was generally believed that the discrepancy lay in the assumption of single scattering, and that second (and higher) order scattering needed to be taken into account.

In [50] Chandrasekhar recalls L. V. King's [2] opinion on the inclusion of multiple scattering in the Rayleigh problem in 1913:

> The complete solution of the problem from this aspect would require us to split up the incident radiation into two components one of which is polarized in the principal plane and the other at right angles to it: the effect of self-illumination would lead to simultaneous integral equations in three variables the solution of which would be much too complicated to be useful.

Chandrasekhar took this as a direct challenge. The solution to the Rayleigh problem became one of the central goals of his theoretical investigations, and his ultimate success with it gave him considerable personal pride.

In order to attack the Rayleigh problem (and equivalent problems involving Thomson scattering), Chandrasekhar first needed a proper framework for the description of polarized radiation in radiative transfer theory. For the special case of scattering with axial symmetry, it was easy to see that the relevant polarization variables could be chosen as the intensities in the two planes parallel and perpendicular to the meridian plane through the direction of the ray, I_l and I_r.[9] In RE-X [19] Chandrasekhar used these

---

[9]The subscripts here are apparently mnemonics using the last letters of "parallel" and "perpendicular."

variables to formulate the radiative equilibrium problem for Thomson scattering and was able to give a complete solution using the discrete ordinate method. One simple, noteworthy result of his numerical calculations was that the degree of polarization of the emergent intensity varied from zero for the normal direction $\mu = 1$ to a maximum of 11.7% for grazing emergence $\mu = 0$. This result is widely quoted as a bound for the degree of polarization that can result from Thomson scattering in such situations.

Chandrasekhar realized that the symmetry arguments that led to a choice of polarization variables in the axially symmetric case were insufficient to deal with the full Rayleigh problem, where the generally oblique angle of the incident sunlight would spoil axial symmetry, so that the principal planes of polarization would not generally be aligned with the meridian plane. He needed a more general formulation for polarization, but searched in vain through the then standard texts. In [50] he recounts his state of mind at this point:

> Nevertheless, it did not seem to me that the basic question could have been overlooked by the great masters of the nineteenth century.

He set his attention on the old papers of Rayleigh and Stokes and quickly found that the required formalism had been developed by Stokes [1] in 1852, almost a century before. This work had fallen into obscurity, but Chandrasekhar seized upon it and resurrected it into the modern form we now know.

Stokes had considered what general description of the polarization of a beam of radiation was required to account for all the various possibilities of linear, circular, or elliptical polarization *and* the possibility that the beam might be completely or partially unpolarized. His conclusion was that such a description required four parameters, which he called A, B, C, and D. Chandrasekhar called these the *Stokes parameters* and gave them the names, I, Q, U, and V, which are now widely used in astronomy. Roughly speaking, I measures the total intensity, Q and U measure the orientation and degree of linear polarization, and V measures the degree of circular polarization.

In a series of papers ([20], [23], and [12]) Chandrasekhar formulated the full scattering problem with polarization. His development of the Stokes parameters is separated into two stages: in RE-XI [20] Chandrasekhar introduced only the theory necessary for the strict Rayleigh problem, without the circularity parameter V; in RE-XV [25] he gave the full theory of all four Stokes parameters. His clear account of the Stokes parameters in RT (section 15) remains one of the best standard treatments of this theory.

In RT (section 17, eq. 226) Chandrasekhar shows that the appropriate transfer equation for polarized transfer is exactly analogous to equation

(4.2) for unpolarized radiation if one regards I as a vector of Stokes parameters and p as an appropriate phase matrix. With the proper description of polarization in hand, Chandrasekhar was able to attack more complex problems of Rayleigh scattering for polarized radiation (RT, chapter 10). In this case the fourth Stokes parameter V plays only a limited role; Chandrasekhar shows that it obeys its own transfer equation uncoupled from the others. One consequence of this is that if the sources of radiation (internal or through boundary conditions) do not introduce any circularity, then the solution will also be completely free of circularity. Thus, both the radiative equilibrium problem and the reflection and transmission problems with incident unpolarized light do not involve V at all.

In RT, sections 69–70, Chandrasekhar gives the solution for the diffuse reflection of a semi-infinite Rayleigh scattering atmosphere, which requires the full non-axially symmetric theory. This solution depends on five H-functions, called H_l, H_r, $H^{(1)}$, $H^{(2)}$, and $H_v(\mu)$, defined by their respective characteristic functions

$$
\begin{aligned}
\Psi_l(\mu) &= \frac{3}{4}(1 - \mu^2), \\[6pt]
\Psi_r(\mu) &= \frac{3}{8}(1 - \mu^2), \\[6pt]
\Psi^{(1)}(\mu) &= \frac{3}{8}(1 - \mu^2)(1 + 2\mu^2), \\[6pt]
\Psi^{(2)}(\mu) &= \frac{3}{16}(1 + \mu^2)^2, \\[6pt]
\Psi_v(\mu) &= \frac{3}{4}\mu^2.
\end{aligned}
\tag{4.25}
$$

For the finite problem there is a corresponding solution for the diffuse reflection and transmission functions in terms of the analogous five X- and Y-functions defined with the same characteristic functions.

In order to make detailed predictions for the polarization of the sunlit sky, Chandrasekhar also realized that there was possibly an additional physical effect to be taken into account, namely, that some of the radiation striking the earth's surface would be scattered or reflected back into the atmosphere, which Chandrasekhar called the *planetary problem*, to distinguish it from the usual case of no incident radiation, called the *standard problem*. With the assumption of ground reflection obeying Lambert's law (reflected intensities independent of angle), Chandrasekhar found a clever way of reducing the planetary problem to the standard one (RT, section 72.1). Thus he was able to give full solutions for the polarization of the sky as a function of the incident sunlight angle.

The solutions for the sunlit sky showed precisely the character of the observations, in particular the various neutral points described above. In 1954, Chandrasekhar and Elbert published a paper [41] with complete polarization sky maps based on his theory, which showed not only the neutral points on the sun's meridian, but also the complete neutral lines.

All this was attained by completely analytic means, except for the final numerical evaluation of the relevant X- and Y-functions. For Chandrasekhar this represented a tremendous triumph for his methods.

4.6 Perspective

Up to this point, attention has been confined primarily to the bulk of Chandrasekhar's work on radiative transfer theory during 1944–50. A full account of its influence on the field in the succeeding forty-five years would be an inappropriate task for a short review such as this. However, it seems appropriate to end with at least a few examples of Chandrasekhar's continuing influence.

In different ways, the discrete ordinate method was both Chandrasekhar's most transient and his most permanent contribution to the field. After RT, the development of analytical radiative transfer rapidly moved toward full treatment of the angular dependence of the solutions, rather than discrete versions. This could already be seen in Chandrasekhar's own work, where he gradually (but not completely) shifted from the discrete ordinate method to the invariance principles to give him the structure of the solutions. Perhaps the most interesting development in this context was the *singular eigenfunction method*, used in plasma physics by Van Kampen [42], and later applied to transfer theory by Case and others (see, e.g., Case and Zweifel [47]). In a sense, this is the true descendant of the discrete ordinate method, since it also starts by asking for solutions of exponential form, but now confronts the true nature of the continuous angular dependence in the scattering integral.

But if the discrete ordinate method has fallen out of favor for analytical radiative transfer, it remains to this day a very strong component of numerical work in stellar atmospheres and other astrophysical applications, because of its simplicity, accuracy, and adaptability to complex physical situations. In this way, the method continues to serve those seeking practical solutions to real physical problems, as Chandrasekhar himself was.

One of the most surprising long-term implications of the invariance principles has been their generalization and development into an entire mathematical field known as "invariant imbedding" by Bellman and others (see, e.g., [48]). This development was based on the recognition that the invariance principles had converted what was a *boundary value problem* (involving boundary conditions at two or more points) into an *initial value problem* (involving boundary conditions at a single point) through the introduction of the reflection and transmission functions and the integrodifferential equations they satisfy. The invariant imbedding methods generalize this idea of transforming from a boundary value problem to an initial value problem to a much wider class of problems than just radiative transfer, including wave propagation and control theory, among others. This was an important practical advance, since initial value problems are generally much more numerically tractable than boundary value problems.

In the introduction to their work [48], Bellman and Wing, after crediting earlier workers, clearly express the special influence of Chandrasekhar in the development of their methods:

> However, it was not until the gifted mathematical physicist S. Chandrasekhar appeared and published his famous book on radiative transfer [RT] that the authors and others began their intensive and extensive studies of the imbedding methods we shall describe in this book. Chandrasekhar developed an elegant theory of principles of invariance, thus completing and considerably extending the ingenious methods of Ambartsumian.

This quotation speaks of the widespread and lasting influence of Chandrasekhar's work and the awe in which it is held. It also shows the wisdom of his custom of distilling his work in the form of a well-written treatise that is much more than a collection of results and facts. In this way he has been able to keep alive the perspective on the field he worked so hard to gain himself, while communicating to others the "elegance and beauty" he felt about it.

References

[1] Stokes, G. G. 1852, *Trans. Camb. Philos. Soc.*, **9**, 399; also 1901, *Mathematical and Physical Papers of Sir George Stokes* (Cambridge: Cambridge University Press), **3**, 233.

[2] King, L. V. 1913, *Phil. Trans. Roy. Soc. London*, **A212**, 375.

[3] Wiener, N., Hopf, E. 1931, *Berliner Ber. Math. Phys. Klasse*, p. 696.

[4] Hopf, E. 1934, *Mathematical Problems of Radiative Equilibrium*, (Cambridge: Cambridge University Press).

[5] Ambartsumian, V. A. 1943, *C. R. (Doklady), Acad. URSS*, **38**, 257.

[6] Wick, G. C. 1943, *Zs. f. Phys.*, **121**, 702.

[7] Ambartsumian, V. A. 1944, *J. Phys. Acad. Sci. USSR*, **8**, 65.

[8] Chandrasekhar, S. 1944, *Astrophys. J.*, **99**, 180. (RE-I)

[9] Chandrasekhar, S. 1944, *Astrophys. J.*, **100**, 76. (RE-II) [SP2-1]

[10] Chandrasekhar, S. 1944, *Astrophys. J.*, **100**, 117. (RE-III)

[11] Cesco, C. U., Chandrasekhar, S., Sahada, J. 1944, *Astrophys. J.*, **100**, 355. (RE-IV)

[12] Chandrasekhar, S. 1945, *Astrophys. J.*, **101**, 95. (RE-V) [SP2-2]

[13] Cesco, C. U., Chandrasekhar, S., Sahada, J. 1945, *Astrophys. J.*, **101**, 320. (RE-VI)

[14] Chandrasekhar, S. 1945, *Astrophys. J.*, **101**, 328. (RE-VII) [SP2-3]

[15] Chandrasekhar, S. 1945, *Astrophys. J.*, **101**, 348. (RE-VIII) [SP2-4]

[16] Chandrasekhar, S. 1945, *Rev. Mod. Phys.*, **17**, 138. [SP2-20]

[17] Chandrasekhar, S. 1945, *Astrophys. J.*, **102**, 402. [SP2-21]

[18] Chandrasekhar, S. 1946, *Astrophys. J.*, **103**, 165. (RE-IX)

[19] Chandrasekhar, S. 1946, *Astrophys. J.*, **103**, 351. (RE-X) [SP2-7]

[20] Chandrasekhar, S. 1946, *Astrophys. J.*, **104**, 110. (RE-XI) [SP2-8]

[21] Chandrasekhar, S. 1946, *Astrophys. J.*, **104**, 191. (RE-XII) [SP2-5]

[22] Chandrasekhar, S. 1946, *Proc. Camb. Phil. Soc.*, **42**, 250. [SP2-23]

[23] Chandrasekhar, S. 1947, *Astrophys. J.*, **105**, 151. (RE-XIII) [SP2-9]

[24] Chandrasekhar, S. 1947, *Astrophys. J.*, **105**, 164. (RE-XIV) [SP2-12]

[25] Chandrasekhar, S. 1947, *Astrophys. J.*, **105**, 424. (RE-XV) [SP2-10]

[26] Chandrasekhar, S., Breen, F. H. 1947, *Astrophys. J.*, **105**, 435. (RE-XVI)

[27] Chandrasekhar, S. 1947, *Astrophys. J.*, **105**, 441. (RE-XVII) [SP2-13]

[28] Chandrasekhar, S., Breen, F. H. 1947, *Astrophys. J.*, **105**, 461. (RE-XVIII)

[29] Chandrasekhar, S., Breen, F. H. 1947, *Astrophys. J.*, **106**, 143. (RE-XIX)

[30] Chandrasekhar, S. 1947, *Astrophys. J.*, **106**, 145. (RE-XX) [SP2-14]

[31] Chandrasekhar, S. 1947, *Astrophys. J.*, **106**, 152. (RE-XXI) [SP2-6]

[32] Chandrasekhar, S. 1947, *Bull. Amer. Math. Soc.*, **53**, 641. [SP2-24]

[33] Chandrasekhar, S. 1948, *Astrophys. J.*, **107**, 48; *ibid.*, 188. (RE-XXII) [SP2-15 SP2-16]

[34] Chandrasekhar, S., Breen, F. H. 1948, *Astrophys. J.*, **107**, 216. (RE-XXIII)

[35] Chandrasekhar, S. 1948, *Astrophys. J.*, **108**, 92 (includes appendix with F. H. Breen); *ibid.*, **109**, 555. (RE-XXIV)

[36] Chandrasekhar, S. 1948, *Proc. Roy. Soc.*, **A192**, 508. [SP2-22]

[37] Chandrasekhar, S. 1950, *Radiative Transfer* (Oxford: Clarendon Press). Reprinted 1960 (New York: Dover). (RT)

[38] Chandrasekhar, S. 1951, *Canadian J. Phys.*, **29**, 14. [SP2-17]

[39] Chandrasekhar, S., Elbert, D., Franklin, A. 1952, *Astrophys. J.*, **115**, 244.

[40] Chandrasekhar, S., Elbert, D. 1952, *Astrophys. J.*, **115**, 269.

[41] Chandrasekhar, S., Elbert, D. 1954, *Trans. Amer. Phil. Soc.*, **44**, 643. [SP2-11]

[42] Van Kampen, N. G. 1955, *Physica*, **21**, 949.

[43] Chandrasekhar, S. 1958, *Proc. Nat. Acad. Sci.*, **44**, 933. [SP2-18]

[44] Anselone, P. M. 1958, *Astrophys. J.*, **128**, 124; **130**, 881.

[45] Horak, H. G., Chandrasekhar, S. 1961, *Astrophys. J.*, **134**, 45. [SP2-19]

[46] Mullikin, T. W. 1964, *Astrophys. J.*, **139**, 1267.

[47] Case, K. M., Zweifel, P. F. 1967, *Linear Transport Theory* (Reading, Mass.: Addison-Wesley).

[48] Bellman, R. E., Wing, G. M. 1975, *An Introduction to Invariant Imbedding* (New York: Wiley).

[49] Chandrasekhar, S. 1989, *Selected Papers*, vol. 2, *Radiative Transfer and Negative Ion of Hydrogen*, (Chicago: University of Chicago Press). (SP2)

[50] Chandrasekhar, S. 1989, Radiative transfer: A personal account, in *Principles of Invariance*, Ed. M. A. Mnatsakanian and H. V. Pickichian (Yerevan: Academy of Sciences of Armenia), 19. [SP2-25]

[51] Wali, K. C. 1991, *Chandra: A Biography of S. Chandrasekhar* (Chicago: University of Chicago Press).

5

The Negative Ion of Hydrogen

A. R. P. Rau

5.1 Prologue

This is primarily the story of a negative ion, offered as a tribute to the memory of a great astrophysicist who contributed to its early understanding. It is also a personal account because both the negative ion, H^-, and the astrophysicist, S. Chandrasekhar, have had major influences on my own career in physics. I begin, therefore, on that personal note.

Just as for many an Indian student over the past several decades, Prof. Chandrasekhar's immense contributions to physics and astrophysics and his stature in the scientific world were an inspiring attraction and motivation as I made my own early decision to enter the field. I first saw and heard him when he delivered a lecture at the University of Delhi in the early 1960s while I was an undergraduate student there. Later, in my decision on a graduate school, his presence on the faculty was one of the determinants in my choosing the University of Chicago. He was just beginning his work in the field of general relativity and, although I chose to work for my thesis in another field and under another faculty advisor, I took almost every course that he taught during my graduate school years. However, his generally stern and serious manner meant that our interactions were confined to the classroom.

Only in my last year at Chicago did I go further, daring to seek time for conversations in his office on topics in my research having to do with H^- that I felt sure would interest him. That step set the stage for what became my standard practice in all subsequent visits to Chicago over the next 12–15 years, to call on him at his office. Very conscious of his dislike of small talk and of his intense focus, I always felt the need to go in with some scientific item to discuss; but, depending on the mood and the interchange, these conversations ranged broadly over physics and philosophy, with his ideas and opinions about them and about the physicists he had known. On many of these occasions he was in a relaxed mood, wanting to talk about some subject, sometimes for long periods. I can still see his smile and hear his characteristic "consequently" as he made a point. These conversations remain among my most memorable experiences in my life in physics. I note one of possible general interest. I had become fascinated with the principle of invariant imbedding and had worked on it for problems in atomic scattering. Since he and Ambartsumian had launched the subject of what he called "the principles of invariance," I went in on one of these visits to tell him about some of my results. During that conversation, he said that of all the pieces of work he had done, many were much more difficult and complicated but the one that "gave the greatest satisfaction" was his work on the principles of invariance; that he found during the course of that work that the subject had "its own natural flow, each step following inevitably from the previous one, nothing forced."

Chandrasekhar's presence there led me to the University of Chicago although not to doing research under his supervision. But I fell in with H^- and am still involved in understanding this two-electron system more than twenty-five years later. Although a simple system, indeed the simplest nontrivial problem in the study of atoms and ions, and one that has been studied since the earliest days of quantum mechanics, it continues to pose challenges to our theoretical understanding. The story of H^- is rich in physics and is far from being closed. It is the prototypical three-body system in atoms and, therefore, the system of choice for studying the intricacies of three-body dynamics in a quantum system. With long-range Coulomb interactions between all three pairs of particles, the dynamics is particularly subtle in a range of energies that lie roughly 2–3 eV on either side of the threshold for break-up into proton + electron + electron at infinity. In this energy range, there is a delicate balance between the attractive and repulsive interactions and, given the low kinetic energies involved, the particles develop strong correlations in energy, angle, and spin degrees of freedom. Perturbation and other conventional techniques fail, posing a challenge to our mathematical and physical understanding. At the same time, such understanding can be expected to apply broadly to all correlations in multi-electron atoms and elsewhere in physics as well. Just as the hydrogen atom is not only the prototype of all one-electron atomic physics but lends its basic ideas, notation, and terminology to other realms, whether excitons and heterostructures or quarkonium, so also will the understanding of H^- apply to and become part of our intuition about coupled, non-perturbative, strongly correlated systems throughout physics.

Further, H^- has been important in the study of our atmosphere (particularly, the ionosphere's D-layer) and, even more, of the atmosphere of the Sun and other stars, as first documented by Chandrasekhar. It has also been central to the development of accelerators, being the ion of choice to start with even when one is interested down the line in beams of protons, mesons, or neutrinos.

5.2 Early history: the ground state

Current research on the strong electron-electron correlations displayed by H^- concentrates on the energy range of high excitation of both electrons, but the early focus, both in physics and in astrophysics, was on the ground state of this negative ion. Indeed, H^- is peculiar in that, unlike the case of other members of the two-electron isoelectronic sequence, correlations between the electrons are crucially important from the start, even in the ground state. This is not surprising because He, Li^+, and other members of this sequence have a dominant Coulomb attraction by the nucleus for both electrons, the interaction between the electrons being perturbative in comparison, so that perturbation and variational methods suffice to give a good description of the binding energy of the ground and low-lying excited states. In H^-, on the other hand, this is no longer true. As is well known from the earliest days, a simple Hartree self-consistent field treatment, with each of the 1s electrons seeing an effective charge, $Z_{eff} = Z - 5/16$, gives a very accurate value of the ground state energy for all $Z \geq 2$ but fails completely for H^-. The predicted variational energy of $-(11/16)^2$ atomic units (1 a.u. $= 27.21\,\text{eV}$), lies above the ground-state Bohr energy of $-1/2$ for H, so that one cannot even conclude that H^- is a bound entity relative to (H + electron at infinity).

It was not till Bethe's 1929 paper, therefore, that there was unambiguous proof of the existence of H^- as a bound system (Bethe 1929). Using the Hylleraas variational wave functions, which involve the three coordinates $s \equiv r_1 + r_2$, $t \equiv r_1 - r_2$, and $u \equiv r_{12}$, Bethe employed a three-parameter $\{\alpha, \beta, k\}$ function of the form $(1 + \alpha u + \beta t^2) \exp(-ks)$ to conclude for the first time that the resulting Rayleigh-Ritz upper bound on the energy lies below $-1/2$ a.u. The presence of the term in αu, involving explicitly the interelectronic distance, speaks to the necessity of including correlations to arrive at such a conclusion. Soon after, Hylleraas, who had pioneered similar calculations for He and higher members of the two-electron sequence, also arrived independently at the same conclusion, his six-parameter calculations giving of course a slightly lower energy (Hylleraas 1930). Today, later generations of such calculations, most notably by Pekeris, Kinoshita, and the others who have followed them (cf. Koga and Matsui 1993; Koga and Morishita 1995), employ hundreds of parameters to give the binding energy of H^- to many-decimal accuracy. These variational calculations have even become a canonical test of new numerical procedures and of the speed and capacity of new generations of computers, the value of (approx.) 0.75 eV for this binding energy being a number remembered by most atomic physicists and astrophysicists.

While many-parameter variational calculations give the ground state energy of H^- to great accuracy, the best experimental values come from a high resolution (0.03 cm^{-1}) laboratory photodetachment experiment with lasers. Extrapolating with the use of the known threshold behavior for this detachment (to be discussed further below in section 5.2.1), the threshold and thereby the electron affinity (or binding energy) has been determined to be 6082.99 ± 0.15 cm^{-1} for H($F = 0$) and 6086.2 ± 0.6 cm^{-1} for the similar D($F = 1/2$) states (Lykke *et al.* 1991).

5.2.1 Opacity of stellar atmospheres

The astrophysical importance of the existence of a weakly bound H$^-$ was first recorded in the literature by Wildt (1939, 1941). The abundant presence of both hydrogen and low energy electrons in the ionized atmospheres of the Sun and other stars argues for the formation of H$^-$ by electron attachment. At the same time, subsequent photodetachment back to H + electron for photon energies larger than 0.75 eV points to its importance for the opacity of these atmospheres to the passage of electromagnetic radiation. Indeed, since most neutral atoms and positive ions have their first absorption at 4 or 5 eV if not larger, H$^-$ is the dominant contributor to the absorption of 0.75–4 eV photons, a critical range of infrared and visible wavelengths. At this point, Chandrasekhar played a crucial role in the subsequent story of H$^-$, both in physics and astrophysics. The continuum absorption coefficient in the solar atmosphere as a function of wavelength was well known. As shown in figure 31 of his book (Chandrasekhar 1960), it increases by a factor of two from 4,000 to 9,000 Å, then decreases to a minimum at 16,000 Å, followed by another increase. It was also known that this shape is characteristic of many other stars with surface temperatures less than 10,000°K. For a long time, until Wildt's suggestion, people had tried to explain this shape as due to continuous absorption by H or some of the other abundant species such as Na, Mg, Ca, Fe, and Si, but the wavelength dependence did not match. Following Wildt, it became natural to look to H$^-$ for the explanation, but attempts and calculated absorption coefficients by Jen, Massey, and Bates, Williamson, Wheeler, and Wildt, etc., were unsatisfactory, as pointed out by Chandrasekhar.

Figure 5.1, drawn from Bethe and Salpeter (1977), shows the very different shapes of the photoabsorption of a negative ion as compared to neutral atoms. Simple physics underlies this difference. Following photoabsorption by H$^-$, the photoelectron departs in a *p*-wave. Just above threshold, the low energy electrons see the angular momentum barrier as the longest-range potential and their tunneling through the barrier suppresses the cross-section. Therefore, as seen in the figure, the cross-section rises from zero, following the Wigner $E^{l+1/2} = E^{3/2}$ law (Wigner 1948). It later rises to a peak value and then gradually falls off. In contrast, photoionization of neutral atoms behaves quite differently. The longest-range potential for the photoelectron is now an attractive Coulomb one which has

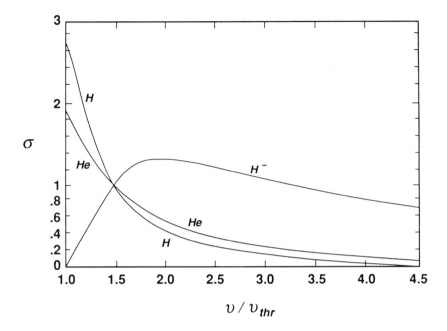

Figure 5.1: Contrast between the cross-section for photoionization of neutral atoms and photodetachment of a negative ion. Horizontal axis is the incident photon frequency in units of the threshold frequency (from Bethe and Salpeter 1997).

two effects. It enhances the wave function near the origin and it also renders l irrelevant since the angular momentum $1/r^2$ potential falls off faster than the Coulomb. The net effect is that photoionization cross-sections are independent of the l-value of the photoelectron and start at a finite value at threshold, to fall off in some fashion at higher energies (cf. Rau 1984a). It is the difference in shapes and, in particular, the feature of a broad region of absorption somewhat above threshold that makes H⁻ important for stellar atmospheres in the 4,000–20,000 Å wavelength range. Next, for quantitative treatment, Chandrasekhar appreciated the special features associated with its weak binding.

5.2.2 Compensating for errors introduced by a diffuse wave function

It is now common wisdom that Rayleigh-Ritz variational calculations may provide accurate energies while the wave function itself may be seriously deficient in other regards. This is particularly important for a weakly bound system such as H^- with its very extended wave function. Together with the deuteron, H^- is a canonical example of a loosely bound quantum system wherein the wave function and probability amplitude can extend beyond the range of the binding potential itself. Chandrasekhar and Krogdahl (1943) pointed out that for the matrix element in the photoabsorption coefficient, the wave function at distances of 4–5 a_0 (Bohr radius, $\simeq 0.53$ Å) is involved and even a many-parameter function may give a poor description at this distance while providing a reasonable energy (which arises more from the wave function at smaller r). As a result, they argued that trial functions be subjected to sum rule tests as indicators of their reliability. Together with the Thomas-Reiche-Kuhn total oscillator strength sum rule, they developed another which related the integrated continuum absorption coefficient to the matrix element of r^2, this sum rule following from the assumption that H^- has only one bound state. This too is an interesting element of the H^- story. For one-electron excitations, a negative ion is very different from its higher isoelectronic analogs. Unlike the infinite number of bound excited states in positive ions and neutral atoms, negative ions have a much sparser spectrum. A rigorous proof that H^- has only one bound state, the ground state, and no singly excited states at all is very recent (Hill 1977 a,b), but not unexpected from the earliest days. Chandrasekhar thanks E. Teller for a conversation regarding his assumption that there are no excited bound states.

Based on these sum rules and under Chandrasekhar's influence, Henrich (1944) did an 11-parameter Hylleraas calculation for H^-. The next important step was taken a year later when Chandrasekhar, pursuing the same theme that the wave functions may be poorer at large r, developed alternative forms of the photoabsorption matrix element (Chandrasekhar 1945). Today these have become standard in our thinking, but they first appeared in this context, with Chandrasekhar pointing out that the usual "dipole-length" form of this electromagnetic coupling, $e\vec{\varepsilon} \cdot \vec{r}$, weights large values of r, precisely where the H^- functions are deficient. But through commutation relationships involving the Hamiltonian, an alternative can be developed which involves matrix elements of the momentum operator p. Another alternative gives an "acceleration" form, and even more alternatives are of course possible. The "momentum" form (today more often called "dipole-velocity") weights the same small-r regions which contribute most to the energy. The wave function in that region being, therefore, expected to be more reliable, the "momentum" calculation of the absorption coefficient may by the same token be more trustworthy. In this vein, Chandrasekhar and Breen (1946), working with the Henrich 11-parameter function, showed that the H^- photoabsorption does indeed peak at 8,500 Å and that H^- can itself account for the continuum absorption coefficient in the solar atmosphere over the entire range from 4,000–25,000 Å [as in figures 3 and 4 of Chandrasekhar and Breen (1946) and figure 32 of Chandrasekhar (1960)]. Characteristic of Chandrasekhar's work at other times and on other problems, they presented extensive tables and, in a succeeding paper with Münch (Chandrasekhar and Münch 1946), applications were made to all A0–G0 stars. (As an aside, there is no general rule that the velocity form of the dipole matrix element is always superior. Indeed, one of my first conversations with Chandrasekhar grew from my analysis (Rau and Fano 1967) that the asymptotic form of a transition matrix element at high momentum transfer was better described by the length form).

5.2.3 Radial correlation and a simple wave function

The above works are notable for their contribution to atomic physics of alternative forms of the photoabsorption matrix element and to astrophysics of a complete accounting of the opacity of stellar atmospheres. For these detailed quantitative applications, Chandrasekhar used many-parameter Hylleraas functions (as also in later work (Chandrasekhar and Herzberg 1955) on He, Li^+ and O^{6+}), but another paper of Chandrasekhar's around this time is notable for a further important insight into the structure of H^-. He introduced a two-parameter trial wave function

$$\exp\left(-\alpha r_1 - \beta r_2\right) + \exp\left(-\alpha r_2 - \beta r_1\right) \tag{5.1}$$

and showed that the energy minimum at $\alpha = 1.03925$ and $\beta = 0.28309$ is sufficient to provide binding for H^- (Chandrasekhar 1944). There is no explicit use in (5.1) of the electron-electron correlation, only the imposition of the Pauli symmetrization requirement for this ^1S function which differs from the Hartree or Hartree-Fock one-parameter wave function wherein $\alpha = \beta = Z_{\rm eff}$. The function in (5.1) is, therefore, referred to sometimes as an "unrestricted" Hartree-Fock function, with the two $1s$ electrons not restricted to see the same effective charge and, therefore, to have the same orbital. The function exhibits a radial "in-out" correlation between the electrons such that when one electron is "in" close to the nucleus, the other is kept "out." Particularly striking is the feature that α is larger than 1, so that the presence of the second, "outer" electron pushes the inner one closer to the nucleus than it would be were it alone bound to the proton. Thereby it "sees" an effective charge larger than the value unity of the proton's real charge! At the same time, the outer electron also sees enough of an effective charge, albeit small, to be itself bound. In the same paper, Chandrasekhar also considered a second function which included an additional factor $(1 + cr_{12})$ in (5.1), which, of course, improved the binding energy and indeed was superior to the 3-parameter Bethe-Hylleraas result. See Bethe and Salpeter (1977) for other discussions on the ground state of two-electron atoms and ions. For recent reviews on negative ions, see Bates (1990) and Buckman and Clark (1994).

The Chandrasekhar function (5.1) shows the specific nature of electron-electron correlation in the ground state of H^-. Of the two kinds of correlations, "angular" between the directions $\hat{r}_1$ and $\hat{r}_2$ and "radial" between the magnitudes r_1 and r_2, it is the latter that proves crucial. Further, the two electrons are on a very different footing, one bound much closer to the nucleus than the other, which is weakly held at a distance $\simeq$ 4–5a_0 from the nucleus. This suggests a very useful next step, of regarding photoabsorption and other collision processes as primarily due to this electron, so that a "one-electron picture" of H^- suffices, this electron regarded as weakly bound in a short-range attractive potential well. An extreme model takes the attraction to be of "zero range" (or, indeed, as a delta-function well) so that a single parameter, the binding energy, characterizes the form of the wave function of the outer electron as $\exp(-k_B r)/r$, where $k_B^2/2$ is the electron affinity of H^- ($\simeq 0.75$ eV). Together with a constant C that allows for normalization, and takes the numerical value 0.31552, Ohmura and Ohmura (1960) took the resulting two-electron wave function

$$(1/2)^{1/2} \left(1 + P_{12}\right) \Psi_0 \left(r_1\right) C \exp\left(-k_B r_2\right)/r_2, \qquad (5.2)$$

where P_{12} is the particle interchange operator, and evaluated the continuum photoabsorption coefficient shown in figure 5.2. This calculation, which describes the outgoing photoelectron by a free p-wave, is extremely simple and completely analytical, while giving a very good description of the absorption coefficient over the entire range from 4,000 to 16,000 Å (section 7.2.4 of Fano and Rau 1986). Figure 5.2 is a re-rendering of the H^- curve in figure 5.1 and similar to equivalent figures in Chandrasekhar's papers that were referred to in section 5.2.2.

Besides photodetachment, collision processes involving H^- are also important in stellar atmospheres. Prominent among these are collisions with the neutral hydrogen and protons that are abundantly present. One among the many results is that "charge exchange," $H^- + H^+ \rightarrow H(1s) + H(nl)$, dominates over "associative detachment," $H^- + H \rightarrow H_2 + e$, in atmospheres with temperatures greater than $8,000°$K and in lower temperature stars with lower surface gravity (cf. Praderie 1971).

5.3   One-electron excitations

The very weak binding already of the ground state, and the absence of a long-range Coulomb attraction for the outer electron, make the excitation spectrum of H^- very different from He and higher members of the isoelectronic sequence. As already noted, there are no other bound states at all (except for an item to be mentioned in section 5.4.5), no counterparts of the multiple infinity of Rydberg states in He. The only excited states are, therefore, of the one-electron continuum of (H + electron) which begins 0.75 eV above the H^- ground state. These are, of course, the same

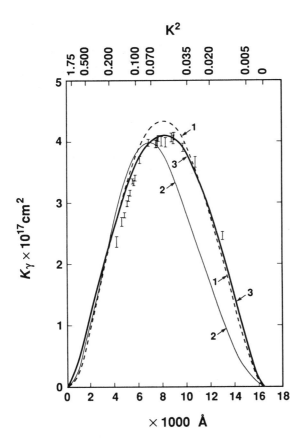

Figure 5.2: Continuous photoabsorption coefficient of H^- from Ohmura and Ohmura (1960). Data from experiment compared with the analytical expression from the simple "zero-range wave function" in (5.2), plotted as curve 3, and with numerical results using a 20-parameter Hylleraas variational wave function in curves 1 ("dipole velocity" form) and 2 ("dipole length" form). Horizontal axes are in photon wavelength (bottom) and photoelectron energy in Ry (top).

states involved in elastic scattering of low energy (< 10 eV) electrons from ground state hydrogen. They have also played an important role as tests of our understanding of electron-atom scattering; see, for instance, a recent review on electron-H scattering (Bray and Stelbovics 1995). Proper treatment of exchange, the singlet and triplet scattering lengths being very different, and of the polarizability of the H ground state (which leads to an attractive r^{-4} potential seen by the scattered electron), are important in describing this scattering. An early variational calculation by Schwartz (1961), with over fifty parameters in the wave function, continues to serve as a benchmark against which later calculational techniques and numerical values are sized up.

As noted in section 5.2.2, although it was suspected from the beginning that there may be no excited $1snl$ bound states of H^-, nevertheless there were temptations from time to time to attribute unexplained diffuse interstellar lines to such states. In the 1950s and 1960s, such suggestions were made with regard to absorption lines at 4,430, 4,760, 4,890 and 6,180 Å. Given the persistence of these speculations and their astrophysical significance, a careful laboratory laser photodetachment experiment by Herbst *et al.* (1974) finally demonstrated conclusively that there is no structure in the detachment cross-section at these wavelengths. As already noted, today we have a completely rigorous mathematical proof (Hill 1977a,b) that there are no one-electron excited bound states of H^-.

5.3.1   *Excitation in static electric and magnetic fields*

With the interaction of photons below 10 eV with H^- essentially understood, more recent work has turned to the effect of additional external electromagnetic fields. Considering first static fields, both electric and magnetic fields are of interest. In both situations, for laboratory field strengths of interest, the couplings $e\vec{\varepsilon}\cdot\vec{r}$ and $(e^2 B^2/8mc^2)r^2 \sin^2\theta$ (from the quadratic or diamagnetic coupling) share the common feature of being negligible at small r and increasing with radial distance. To an excellent approximation, therefore, they affect only the final state of the detached photoelectron, and that too only at large distances. The initial state of H^- and the initial absorption of the photon by it are essentially unchanged from the zero-field case. Only the large-r wave function of the photoelectron has to be recast in terms of the eigenstates of a free electron in the external potential, and these are well known, Airy functions and Landau functions respectively. The cross-section for photodetachment in the presence of the external field F takes, therefore, a simple form

$$\sigma(F) = \sigma(F = 0)\mathrm{H}_F(k), \qquad (5.3)$$

where $\mathrm{H}_F(k)$ is a "modulation factor" which depends on the outgoing electron's momentum k. Simple analytical expressions for this factor have been developed (Greene 1987; Rau and Wong 1988; Du and Delos 1988). In the case of an external static electric field, experimental data are available (Bryant *et al.* 1987) for fields of 10^5 V/cm and, as shown in figure 5.3, are in excellent agreement with theoretical calculations (Rau and Wong 1988). Note that photoabsorption sets in below the zero-field threshold energy of 0.75 eV because of field-assisted tunneling, and that above this energy the cross-section displays oscillations about the zero-field cross-section.

It is worth taking note of the experiment that gave the results in fig. 5.3. Taking advantage of the fact that most particle physics accelerators that ac-

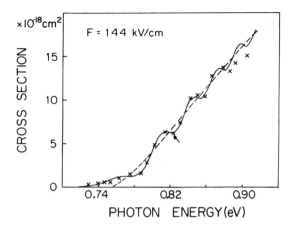

Figure 5.3: Photodetachment of H^- just above the detachment threshold at 0.75 eV (dashed line) and in the presence of a strong electric field F (solid line). Note in the latter that absorption sets in "below threshold" because of field-assisted tunneling through the sloping potential cFx and that field-induced modulations appear at higher energies. The arrow marks the data point that was normalized to the theoretical expression (from Rau and Wong 1988).

celerate protons start with H^- as the initial species (because it has the same mass and magnitude of the charge and the two electrons can be stripped off after acceleration), a group has conducted a series of detailed studies of H^- over the last twenty years at the LAMPF accelerator in Los Alamos (see a review: Bryant *et al.* 1981). The relativistic nature of the 800 MeV H^- beam has been cleverly exploited to make certain measurements that are not otherwise easily carried out. Thus, in the context of figure 5.3, such large electric fields were realized by imposing a modest magnetic field of 10^3 gauss on the H^- beam as it was photodetached by a laser (Bryant *et al.* 1987). This laboratory magnetic field transforms in the rest frame of the H^- beam into both a comparable magnetic field and a large electric field.

5.3.2 Multiphoton detachment

In recent years, intense lasers have made it possible to observe effects of multiphoton absorption by atoms and ions. Among notable effects is "above-threshold ionization," wherein more photons are absorbed than are necessary to break up the system, the extra energy going to increase the kinetic energy of the ejected photoelectron by multiples of the photon energy. The Los Alamos H^- beam mentioned above in section 5.3.1 has also been used for similar studies of "above-threshold detachment" (Tang *et al.* 1989,

1991). With a CO_2 laser beam of laboratory photon energy 0.117 eV, the relativistic Doppler shift makes it appear in the frame of H^- as of energy 0.08–0.39 eV, depending on the angle between the photon momentum and the H^- beam. Detachment, which requires 0.75 eV, takes place therefore as the result of the absorption of two to nine photons or more. Thereby multiphoton detachment and above-threshold detachment have now been experimentally studied in H^- (fig. 5.4). Theoretically, this problem had been investigated earlier and by several groups, again primarily because the negative ion is a simpler system than an atom with its Coulomb field and spectrum (cf. Crance and Aymar 1985; Arrighini *et al.* 1987; Geltman 1990 and 1991; Liu *et al.* 1992; Laughlin and Chu 1993; and the collection Gavrila 1992). The coupling of an intense time-dependent electromagnetic field to an atom is still an unsolved problem because neither the external field nor the internal binding field can be treated perturbatively. With a negative ion, especially in a zero-range description as in section 5.2.3, the internal field is at least simple, and, therefore, H^- has been a system of choice for the study of such time-dependent problems (Gavrila 1992; Wang and Starace 1993).

5.4   Two-electron excitations

Although H^- is very different from He as regards the one-electron excitation spectrum, they are on the same footing when it comes to states in which both electrons are excited. Indeed, upon regarding them as two excited electrons around a positive charge, H^-, He, Li^+, ..., are exact "iso-double-electronic" analogs, differing only in the magnitude of Z, the central positive charge, $Z = 1, 2, 3, \ldots$. Therefore, as a prototype for the study of doubly-excited atomic states, H^- is as good a candidate as He. Figure 5.5 is a sketch of the entire spectrum of H^- for $L = S = J = 0$, that is, $^1S^e$ states. Similar sketches describe states of other L, S, and J. Figure 5.5(a) provides the conventional independent-electron labeling, each group of states described as $H(N \geq 2) + e(n \geq N)$. Note N Rydberg series of states (described as l^2 with $l = 0, 1, \ldots, N-1$) below each single ionization limit, $H(N)+$ electron at infinity, along with their associated one-electron continuum above this limit. The first group below $N = 2$ lies, therefore, in the vicinity of 10.95 eV above the ground state of H^-. For analogies to quark families, see Rau (1992).

These states in which both electrons are excited out of the ground $1s$ orbitals are not strict bound states, even in the absence of coupling to the radiation field. This is clear from figure 5.5, since these states are degenerate with one or more continuum states of electron plus H (lower N). Thus, the $1/r_{12}$ interaction itself mixes all these degenerate states, and the physical eigenstates of the Hamiltonian are superpositions of both, with bound and continuum character. They are quasi-bound, "autoionizing" states

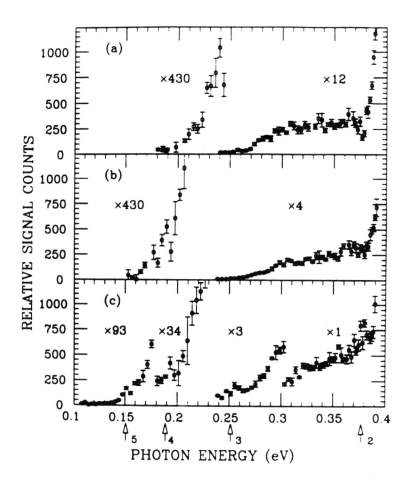

Figure 5.4: Multiphoton detachment of H^- at different laser intensities: (a) $4\,GW/cm^2$, (b) $6\,GW/cm^2$, and (c) $12\,GW/cm^2$. Multiplication factors indicate the magnification in signal counts (from Tang *et al.* 1991).

(cf. sections 10.1 and 10.2 of Fano and Rau 1986). In a descriptive picture, were one to excite both electrons in H^- to one of the states in figure 5.5(a), one electron could drop back into a bound state of hydrogen with lower N, the other then ending up with that released energy which is sufficient to let it escape to infinity ("autoionization"). These states manifest themselves as resonances in electron-hydrogen scattering, as, for instance, in the elastic scattering of 10.2 eV electrons from the ground state of hydrogen. They are also seen in photodetachment of H^- in a corresponding energy range. This range being approximately 10.95–14.35 eV, energies not easily avail-

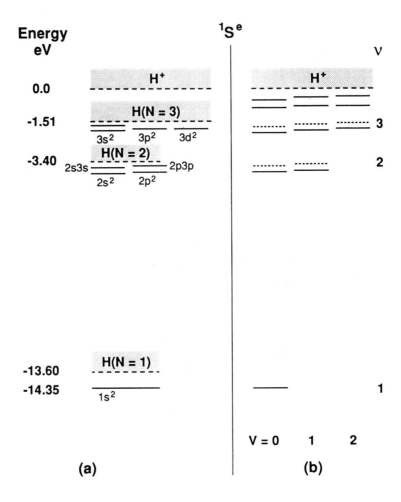

Figure 5.5: Spectrum of ^1S states of H$^-$. (a) In independent-electron labeling with Rydberg series $|Nl, nl\rangle$ converging to states of H(N) plus one electron at infinity. Note the single bound state $1s^2$ of H$^-$ but a rich spectrum of doubly-excited states, with both N and n larger than 1. (b) In a pair labeling, with principal quantum number ν, angular correlation number v, and radial correlation number $\eta(= 0, - ; = 1, - ;$ etc.). States with same v and η form "pair-Rydberg" series converging to the double-detachment threshold of H$^+ + (ee)$ pair at infinity.

able with laboratory lasers, it was once again the Los Alamos experiment that provided laboratory studies of these doubly-excited states.

5.4.1 Experimental observation

One of the clever exploitations of a relativistic H^- beam is to use the Doppler effect to tune the laser frequency as seen in the frame of the H^-. By changing the angle α between the two beams, a laboratory frequency ν_0 appears to the H^- as $\nu = \gamma\nu_0(1 + \beta\cos\alpha)$, where $\beta \equiv v/c$ and $\gamma \equiv (1 - \beta^2)^{-1/2}$. In the Los Alamos experiments, ν/ν_0 could be adjusted from 0.293 to 3.413. As a result, the fourth harmonic of a YAG laser with $h\nu_0 = 4.66$ eV can be used to cover the region of doubly-excited states in H^- from 10 to 15 eV (Bryant *et al.* 1981). Small changes in α provide for the tuning, allowing experimental resolution (mainly limited by H^- beam stability) of a few meV. It is worth noting the interesting combination of circumstances that speaks eloquently to the unity of physics. An 800 MeV H^- beam at an accelerator built for studies of mesons and neutrinos is used to study details of an atomic state around 10 eV with a few meV accuracy! The resulting photodetachment cross-section is shown in figure 5.6 and represents an extension of figure 5.2 to higher energies (Broad and Reinhardt 1976).

Against a background of the one-electron continuum absorption, doubly-excited states of H^- appear in figure 5.6 as groups of resonances in the vicinity of the various single ionization limits H(N)—see figure 5.5. Selection rules for single photon absorption by the ground $^1S^e$ state lead to $^1P^o$ states, the lowest such doubly-excited states being the ones associated with H(N = 2) and loosely termed $2s2p$. Experimental data in figure 5.7 show in greater detail that there are two prominent resonances, a sharp one just below the N = 2 threshold at 10.95 eV above the ground state and a broader one just above that threshold (Bryant *et al.* 1977). Similar sets of resonances have been resolved in the experiment below higher N up to N = 7, figure 5.8 providing as an example the N = 5 set (Harris *et al.* 1990). Although the integrated oscillator strength over the resonances may be small compared to the background continuum, the resonance structures are dramatic over the narrow energy ranges where they occur. Although they have not been observed or discussed so far in the astrophysical context, the advent of far-ultraviolet telescopes may well make them relevant for future studies of stellar atmospheres in this wavelength range of 900–1,100 Å.

5.4.2 Strong correlations

Once both electrons are excited away from the nucleus, correlations between them become more important. In H^-, as already noted, radial corre-

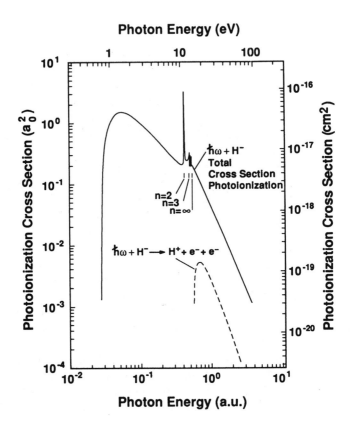

Figure 5.6: Extension of photodetachment cross-section of H^- in figure 5.2 to higher energies, with horizontal axis reversed, photon energy now increasing to the right, and showing doubly-excited states as sharp resonances (from Broad and Reinhardt 1976).

lations are already important in the ground state but, whether in H^- or He, radial and angular correlations are crucial in doubly-excited states (Ho and Callaway 1984, 1986; Pathak *et al.* 1988). The higher the excitation and, therefore, the further removed are the electrons from the central attraction and the slower they get, the more important these correlations become, reaching an extreme near the 14.35 eV energy of the threshold for double break-up. Increasingly, independent particle pictures lose their meaning, as each electron feels as much of a force from the other as it does from the nucleus, and proper understanding requires a treatment of the three-particle system as a whole with a joint description of both electrons—a "pair" of electrons. Pair quantum numbers and even pair coordinates are, therefore, part of the story of H^- in the 10–16 eV energy range (Rau 1984b).

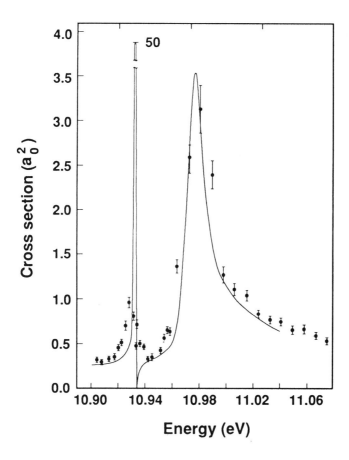

Figure 5.7: Doubly-excited $^1P^\circ$ resonances of H$^-$ in the vicinity of the H(N=2) threshold as observed in the photodetachment cross-section, with a sharp "Feshbach" resonance just below and a broad "shape" resonance just above the 10.95 eV threshold (from Bryant *et al.* 1977).

Angular: A glance at figure 5.5(a) suffices to emphasize that strong angular correlations set in already at N = 2 with the first doubly-excited states. This is because states differing only in l, such as $2s^2$ and $2p^2$, or $2sns$ and $2pnp$, are degenerate in the absence of the electron-electron interaction and, therefore, will be strongly mixed when that interaction is taken into account. Put another way, in the presence of the electric field due to the other electron, the degenerate l states of the hydrogenic manifold of one electron are strongly mixed so that l loses meaning as a quantum number. A first step is to use degenerate perturbation theory within the set of $(nl)^2$ states (for ^1S symmetry, and counterparts for other values of L, S and J)

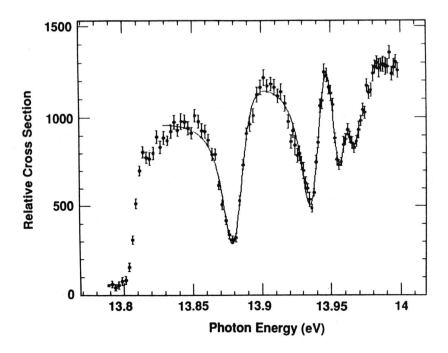

Figure 5.8: $^1P^\circ$ resonances of H^- below the $H(N=5)$ threshold, with data fitted to "Fano resonance profiles" (from Harris *et al.* 1990).

to get new eigenstates that include the $1/r_{12}$ interaction. The label l is thereby replaced by $v = 0, 1, 2, \ldots, (n-1)$, which is the first of such "pair quantum numbers" and which may be regarded as the quantum number associated with θ_{12}, the angle between $\hat{r}_1$ and $\hat{r}_2$.

The mixing coefficients in $|v\rangle = \sum_l \langle l|v\rangle |l\rangle$ can be obtained by numerical diagonalization (Rau and Molina 1989; Rau 1990a,b) and are seen to be very well described by analytical expressions from a group-theoretical model (see a review: Herrick 1983). In this description, the product of the individual SO(4) representations that hold for each electron independently (the well-known symmetry of the hydrogen atom) is reduced to a single SO(4) for the pair to provide simple analytical formulae for $\langle l|v\rangle$. Examination of the corresponding wave functions of the mixed states $|v\rangle$ shows, as may be expected, that the lowest-lying state with $v = 0$ has a concentration at $\theta_{12} \simeq \pi$ which minimizes the electron-electron repulsion, whereas the state at the opposite end with $v = n - 1$ has a concentration around $\theta_{12} = 0$. The former is of greatest interest because it describes the lowest energy states with greatest stability. The concentration at $\theta_{12} \simeq \pi$

in this state is seen to have a width that scales as $n^{-1/2}$. The extreme concentration due to angular correlation, reached as $n \to \infty$, the double-detachment threshold, has the two electrons lying on opposite sides of the nucleus (Rau 1990a). The SO(4) model also applies to $L \neq 0$ states in which the individual l_1 and l_2 values of the electrons need not be equal, leading to two pair quantum numbers, called K and T, which replace l_1 and l_2 (Herrick 1983). The former, simply related to v, is associated with the pair coordinate θ_{12}, whereas T is a measure of $(\vec{L} \cdot \hat{r}_<)^2$, that is, of the projection of the total orbital angular momentum on the radial vector of the inner (N) electron (see, for instance, sections 10.3 and 10.5.2 of Fano and Rau 1986).

Radial: Whereas angular correlations are similar in all doubly-excited states, having their origin in the degenerate l-mixing, under radial correlations two-electron excitations divide into two classes. One, with an "in-out" aspect as in section 5.2.3 for the ground state of H^-, has $r_> \gg r_<$, whereas the second group has comparable radial excitation ($r_1 \simeq r_2$) of the two electrons. The first group may be termed "planetary" (Percival 1977) in that each electron can be ascribed an individual principal quantum number (or orbit) with $n >$ N, these one-electron quantum numbers retaining their meaning (Rau 1984b). Such states in H^- may also be termed "Coulomb-dipole" because the inner electron sees a dominant Coulomb field whereas the outer electron sees the dipole field of the other two particles (this is again the permanent electric dipole moment or the "linear Stark effect" of H(N ≥ 2) arising from the degenerate l-mixing). In He and other isoelectronic analogs, however, such planetary states have both electrons seeing a Coulomb attraction, the outer having in addition a dipole field. The second class of states with a radial correlation such that both electrons share the excitation energy comparably have been called "Wannier ridge" states (Buckman *et al.* 1983), and for them H^- and He are essentially similar. In their description, particularly as one approaches the double-detachment threshold, a true pair picture with no reference to independent particle coordinates (r_1 and r_2) or quantum numbers (N and n) or ionization limits (N) becomes necessary (Rau 1983 and 1984b).

5.4.3 *Hyperspherical coordinates and the two-electron Schrödinger equation*

Over the last twenty-five years, the study of doubly-excited states has prompted the use of a set of coordinates that deal with the pair of electrons from the start (cf. chap. 10 of Fano and Rau 1986; Lin 1986). An alternative, which also treats the three-body system as a whole, sets up a correspondence between H^- and H_2^+, using the language of molecular orbitals and potential wells, r_{12} playing the role of the internuclear distance

R (Feagin and Briggs 1986, 1988; Rost and Briggs 1988; Feagin 1988). Both these approaches have common elements and have been useful in understanding doubly-excited states. I will now turn to the more direct recasting of $\vec{r}_1$ and $\vec{r}_2$ of the atomic system in terms of joint coordinates called hyperspherical coordinates, because this also generalizes immediately to problems involving more electrons. These coordinates were actually first introduced by Bartlett (1937) and Fock (1954) in the study of the ground state of He, to handle the description of the wave function wherein both r_1 and r_2 go to zero, the same considerations applying also of course to H$^-$. The coordinates were also invoked by Wannier (1953) over forty years ago in treating the threshold double escape of two electrons (we will return to this in section 5.5) before they were popularized in recent years for the study of doubly-excited states.

The angle θ_{12} between $\hat{r}_1$ and $\hat{r}_2$ has already been introduced. Next, the radial distances are replaced by the "circular coordinates" in the $r_1 - r_2$ plane, $R \equiv \left(r_1^2 + r_2^2\right)^{1/2}$, $\alpha \equiv \arctan\left(r_2/r_1\right)$. Together with the three Euler angles to describe the position of the (proton + electron + electron) plane in space, the coordinates R, α, and θ_{12} provide a set of six "pair coordinates" to replace $(\vec{r}_1, \vec{r}_2)$ in the independent-particle picture. The three pair coordinates also provide a very natural description of the system, with R providing the overall size and a measure of excitation, α indexing the radial correlation, and θ_{12} the angular correlation between the electrons. For a general three- or N > 3–body system with arbitrary masses for the particles, hyperspherical coordinates can be defined to incorporate the masses so that $R \equiv \left[\sum_i m_i r_i^2 / \sum_i m_i\right]^{1/2}$ is actually the radius of gyration (Fano 1981). The coordinates then are those of a $3(N-1)$-dimensional sphere with R the radius and $\hat{R}$ the hypersurface of such a sphere.

The Schrödinger equation for H$^-$ or its isoelectronic analogs takes the following form in hyperspherical coordinates and in atomic units (cf. section 10.4 of Fano and Rau 1986):

$$\left[-\frac{1}{2}\left\{\frac{d^2}{dR^2} - \frac{\Lambda_{\hat{R}}^2 + 15/4}{R^2}\right\} + \frac{Z(\hat{R})}{R}\right] R^{5/2}\, \Psi\left(\vec{R}\right) = E R^{5/2}\, \Psi\left(\vec{R}\right), \quad (5.4)$$

where $\Lambda_{\hat{R}}^2$ is the Laplacian or "grand angular momentum" operator involving derivatives with respect to α and θ_{12} (the counterpart of $\vec{L}^2$ in three dimensions), and $Z(\hat{R})$ is an "effective charge operator":

$$Z(\hat{R}) = -Z\left(\frac{1}{\cos\alpha} + \frac{1}{\sin\alpha}\right) + \frac{1}{\left(1 - \sin 2\alpha \cos\theta_{12}\right)^{1/2}}. \quad (5.5)$$

Each of the three terms in (5.5) arises from the three pairs of Coulomb interactions in the system once a common dimensional $1/R$ element has

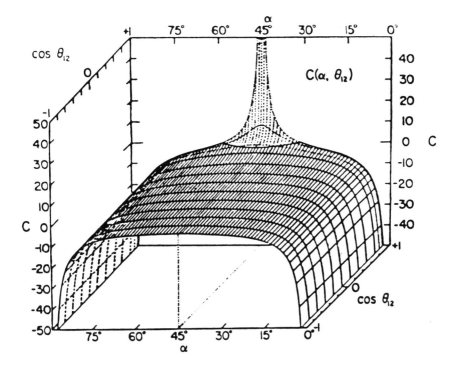

Figure 5.9: Potential surface $C(\alpha, \theta_{12}) = Z(\hat{R})$ in (5.5) for H^- in hyperspherical coordinates (from Lin 1974). Note valleys at $\alpha = 0$ and $\pi/2$, a peak at $\alpha = \pi/4$, $\theta_{12} = 0$, and a saddle point at $\alpha = \pi/4, \theta_{12} = \pi$.

been factored out. The "potential surface" described by $Z(\hat{R})$ in (5.5) is shown in figure 5.9. Note deep valleys at $\alpha = 0$ and $\pi/2$ which correspond to r_2 or r_1 vanishing and an infinite repulsion at $\alpha = \pi/4, \theta_{12} = 0$ which marks the electron-electron repulsive singularity at $\vec{r}_1 = \vec{r}_2$. Only half the potential surface from $\theta_{12} = 0$ to π is shown, the identity of the two electrons repeating a reflected segment of figure 5.9 for $\theta_{12} = \pi$ to 2π. Finally, the surface has another critical point, a saddle at $\alpha = \pi/4, \theta_{12} = \pi$, that is, when $\vec{r}_1 = -\vec{r}_2$. This saddle point will be crucial in the rest of this story.

The H^- Schrödinger equation in hyperspherical coordinates takes the form in (5.4), which is very similar to that of hydrogen in three-dimensional spherical coordinates, with a radial and angular kinetic energy and a Coulomb potential. The major difference is that the charge depends on the hyperangles, making the equation nonseparable. Since other multi-particle problems of atomic and molecular systems also take a form similar to (5.4) with $3(N-1)$-dimensional hyperspherical coordinates (and $5/2$ replaced by $(3N - 4)/2$ and a charge that depends on the $(3N - 7)$ hyperangles, the H^- system takes on added significance as a prototype for such non-separable problems. In particular, the potential surface in figure 5.9 is the simplest prototype of such multi-dimensional potential surfaces, so that the understanding of quantum-mechanical solutions of a configuration point moving on such a surface may be expected to play the same role in shaping our intuition about more general problems of chemical transformation as the Coulomb and harmonic oscillator potentials have played for two-body problems in physics. Besides maxima and minima, there is one saddle point in such a two-variable potential surface which will turn out to be especially interesting for the rest of our story of H^- and also of especial significance generally because saddle points proliferate when more variables are involved.

5.4.4　*Low-lying states—adiabatic treatment*

The Schrödinger equation in (5.4) is nonseparable, so that no exact solutions are feasible. The different scaling in R of the Coulomb potential and angular kinetic energy, together with the dependence of Z on $\hat{R}$, are at the heart of this nonseparability, so that expansion/excitation of the system (in R) is inextricably coupled to the radial and angular correlations (in α and θ_{12} respectively). For the low-lying doubly-excited states, however, that is, with $N \leq 6$ in figure 5.5, an adiabatic separation of R from α and θ_{12} has proved successful (Macek 1968). That is, correlations develop faster than the general expansion of the system under excitation. Much as in the Bohr-Oppenheimer procedure for molecules, with R held fixed, the angular part of the Hamiltonian is solved to provide R-dependent eigenvalues and eigenfunctions:

$$\left[\frac{1}{2}\Lambda_{\hat{R}}^2 + RZ(\hat{R})\right] \Phi_\mu(R; \hat{R}) = R^2 U_\mu(R) \, \Phi_\mu\,(R; \hat{R}). \qquad (5.6)$$

These so-called "channel functions" then provide a basis for expansion of the wave function Ψ in (5.4):

$$\Psi(\vec{R}) = \sum_\mu F_\mu(R) \, \Phi_\mu(R; \hat{R}). \qquad (5.7)$$

The eigenvalues $U_\mu(R)$ appear as potential wells in the resulting coupled radial equations for $F_\mu(R)$, the coupling between μ and μ' provided by matrix elements of d/dR and d^2/dR^2 between Φ_μ and $\Phi_{\mu'}$. The initial adiabatic approximation neglects these couplings, viewing each $F_\mu(R)$ as an eigenfunction of a single potential well $U_\mu(R)$, the resulting eigenvalues being the doubly-excited state energy levels (Lin 1986; chap. 10 of Fano and Rau 1986).

A variety of different approaches may be used to solve (5.6) to get the U_μ and Φ_μ. One is to expand Φ_μ in terms of the basis provided by the "hyperspherical harmonics" that are the eigenfunctions of $\Lambda_{\hat{R}}^2$ and are analogs of the ordinary spherical harmonics (Vilenkin 1968; Avery 1988). Generally, the adiabatic hyperspherical calculations have viewed doubly-excited states as groups converging to H(N) plus electron and, therefore, at large R have imposed the corresponding boundary conditions in defining the channel functions. Figure 5.10 shows the three lowest potential curves in the vicinity of the N = 2 threshold (Lin 1975, 1976). In independent particle terms, $2snp$ and $2pns$ states are strongly mixed in the combinations marked + and −, together with a smaller admixture of the third configuration, $2pnd$, of this $^1P^\circ$ symmetry (Cooper *et al.* 1963). The curve marked − has a long-range attractive tail which corresponds to the $1/r^2$ dipole potential on the outer electron because of the $2s - 2p$ degeneracy of H(N = 2). This well holds an infinite number of "dipole-bound" states, the lowest of which is in very good correspondence with the experimentally observed sharp feature, a "Feshbach resonance", in figure 5.7. The other broader feature in that figure is seen to correspond to the + curve in figure 5.10, which has a broad barrier region for $R > 15a_0$, a state just above the -0.25 a.u. N = 2 threshold energy being temporarily trapped by this barrier to appear as a "shape resonance." For a molecular orbital treatment and labeling of similar potentials to those in figure 5.10, see Feagin (1988).

Adiabatic hyperspherical calculations have been carried out to higher N, the most extensive H$^-$ studies being found in Koyama *et al.* (1989), Sadeghpour and Greene (1990), and Sadeghpour (1991). Figure 5.11 shows a sequence of potential wells and of doubly-excited state Feshbach resonances held in the lowest well, which has an asymptotic attractive potential due to the dipole moment of H(N). These resonances conform well (Sadeghpour 1991) to the experimental data, as illustrated in figure 5.8. In each of these Rydberg series below each H(N) threshold, the lowest level (N = n in independent particle language) fits a "six-dimensional Rydberg formula" (Rau 1983) to be discussed further in section 5.5, while higher members of the series (with $n > N$) then fit the spacings expected of "dipole-bound" states (Gailitis and Damburg 1963; Gailitis 1980). The greatest numerical accuracy at lower N has been achieved by a so-called "diabatic-by-sector" handling of the coupled hyperspherical equations (Tang *et al.* 1992).

5.4.5 Doubly-excited states in an external field

The same Los Alamos experiment that observed the doubly-excited states of H$^-$ shown in figures 5.6–5.8 has also studied the effect of a strong electric field on them (Gram *et al.* 1978; Bryant *et al.* 1983). Of the two N = 2 $^1P^\circ$ resonances in figure 5.7, the sharp Feshbach one with an almost degenerate

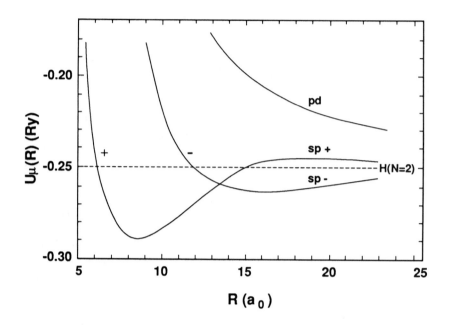

Figure 5.10: Adiabatic potential wells $U_\mu(R)$ in (5.6) for $^1P^o$ states of H$^-$ converging to H(N=2) (from Lin 1975).

$^1S^e$ resonance (Callaway and Rau 1978; Wendoloski and Reinhardt 1978). With increasing strength of the external electric field, the resonance finally disappears, whereas the shape resonance just above threshold persists to even larger fields. Hyperspherical calculations that give the potential barrier in figure 5.10 which accounts for this resonance have been extended to include the effect of the electric field on this barrier and, thereby, on the resonance (Lin 1983; Slonim and Greene 1991; Du et al. 1993). Similar studies have also been carried out for N = 4 and 5 (Zhou and Lin 1992) and also by conventional configuration interaction calculations with independent particle functions (Ho 1995; Bachau and Martin 1996).

Of particular interest among the N = 2 states is the lowest one of $^3P^e$ symmetry, described in independent-electron terms as $2p^2$. This is bound below the H(N = 2) threshold with about 9.6 meV. The only one-electron continuum at this energy being H(N=1) + electron, which cannot form a state with quantum numbers $^3P^e$, this state is forbidden to autoionize. It can only decay into this continuum by also simultaneously radiating a photon along with the electron, these two particles sharing the excess energy

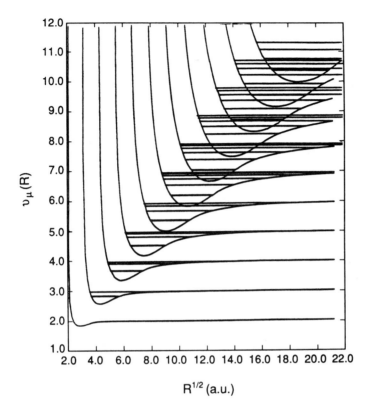

Figure 5.11: Similar to figure 5.10 but showing only the lowest potential well for $^1P^o$ states below successive thresholds H(N). Effective quantum numbers $\nu_\mu(R) = [-U_\mu(R)/13.6 \text{ eV}]^{1/2}$ are plotted as a function of $R^{1/2}$. In each well, the lowest bound states are shown as horizontal lines (from Sadeghpour and Greene 1990).

of $\simeq 10.2$ eV (Drake 1973). The inverse process of radiative attachment, $\text{H(N=1)} + e + \gamma \rightarrow \text{H}^-$ $(^3P^e)$, has been suggested as an efficient absorption mechanism for ultraviolet light (Drake 1974) and has been observed in rocket measurements of Zeta Tauri stars (Heap and Stecher 1974). This very long-lived H$^-$ state has not so far been observed in the laboratory, but double detachment (Mercouris and Nicolaides 1993) and photodetachment to the $2s2p$ $^3P^o$ state (Du et al. 1994) have been studied theoretically as possibly feasible experiments.

5.5   High excitation and the double continuum

As discussed in the previous section, we have now a fairly good and complete understanding of the low-lying doubly-excited states of H$^-$, both experimentally and theoretically, including the nature of radial and angular correlations in them. For states above N = 6, however, and as the double-detachment threshold at 14.35 eV is reached, our knowledge is extremely fragmentary (Nicolaides and Komninos 1987; Pathak *et al.* 1989; Ho 1990, 1992). Experimental data on the series converging to such larger values of N are sparse and show, as also in He in the corresponding energy range, $\simeq 79$ eV (Domke *et al.* 1991, 1995; Wintgen and Delande 1993), that the series overlap and perturb one another. Hyperspherical treatments as in figure 5.11 lead to the same conclusion of many overlapping potential wells (Sadeghpour 1991). Although at first some analysis can be, and has been, carried out to handle interlopers perturbatively, it is clear that this cannot extend to really high N and the double-detachment threshold. This is due to the very nature of a Coulomb potential with its high density of states. In the similar situation of one electron in a three-dimensional Coulomb potential (plus shorter range distortions), quantum defect theory, based on continuity properties in the vicinity of the ionization threshold, affords effective handling of a highly excited electron (chap. 5 of Fano and Rau 1986). Now, for a pair moving in the six-dimensional Coulomb field in (5.4), it is likewise natural to focus on the double-detachment threshold as the starting point on which to base analysis of the higher reaches of the doubly-excited state spectrum (Rau 1984b).

5.5.1   Description as a pair of electrons

The methods described in sections 5.4.3 and 5.4.4, while employing joint or pair coordinates, nevertheless do not give up completely the crutch of independent-particle descriptions. In particular, the use of N, the inner electron's principal quantum number, both in the classification of doubly-excited states and in developing the channel functions Φ_μ to converge to successive single-detachment thresholds, introduces an element foreign to a fully pair treatment. Indeed, calculations revert at large R even to the independent-particle coordinates r_1 and r_2 to get efficient convergence to the single-detachment thresholds (Christensen-Dalsgaard 1984). Only recently have treatments emerged that depart from this, proceeding to large R without ever reverting to single particle aspects (Zhang and Rau 1992; Heim *et al.* 1996). Once double detachment is energetically possible above 14.35 eV, it is also *the pair* that escapes to $R = \infty$, as emphasized in Wannier's treatment of the threshold law for this process, a point of view essentially different from all others that regard *two electrons* as escaping from a central positive charge (Rau 1971, 1984b,c).

The focus on the pair coordinates (R, α, θ_{12}) throughout also organizes the spectrum of doubly-excited states in figure 5.5, the quantum numbers μ, and the potential wells, in an alternative but very different way from the discussion so far. Thus, instead of series associated with each N as in figure 5.5(a), consider the alternative in figure 5.5(b), where the same levels have been redrawn with no reference to N and grouped differently (Rau 1984b). For each v, the pair quantum number that indexes the number of nodes in θ_{12}, levels drawn similarly (solid, dashed, etc., lines) form series converging to the double-detachment threshold. For the system of a pair of electrons in a Coulomb field, this is the only limit compatible with the picture of the system as a whole, shorn of all independent-particle aspects (Read 1982; Rau 1983, 1984b). A pair principal quantum number v, conjugate to R, and a "radial correlation quantum number" η, which counts the nodes in α, provide an alternative basis $|v, \eta, v\rangle$ to the adiabatic hyperspherical basis, $|N, n, v\rangle$, or the independent-particle basis, $|N, n, l\rangle$. The H^- Schrödinger equation does not separate in any coordinate system and, as in any such non-separable problem, any complete basis set affords a description of the whole system. The choice among alternatives is made on the basis of appropriateness, one or the other affording a more economical description depending on the energy range. For the high doubly-excited states and nearby double continuum in the vicinity of 14.35 eV, when many basis states of the other sets are strongly mixed, it is the $|v, \eta, v\rangle$ that more nearly conform to the physical eigenstates.

States with the same v and η but successive values of v form a "six-dimensional" Rydberg series with an appropriate Rydberg formula (Rau 1983) and, as already noted in section 5.4.4, provide a good description of the lowest states below each N. Examination of their wave functions is also instructive. The adiabatic hyperspherical calculations such as in figure 5.11 had already demonstrated (Lin 1986) that the channel functions Φ_μ near $R = R_{\min}$ for the lowest potential well converging to each N showed a concentration of the wave function near $\alpha = \pi/4, \theta_{12} = \pi$, the configuration in which the two electrons are on opposite sides of the nucleus at equal distances and which is the saddle point of figure 5.9. Viewed as a set of potential wells, this concentration passed at larger R from one well to the next of higher N roughly in the vicinity of the prominent avoided crossings seen in figure 5.11. The locus of avoided crossings in such a figure is well described by the value of the potential at the saddle point, namely, $Z(\hat{R} = \text{saddle point})/R$.

This suggests an alternative "diabatic" handling of the hyperspherical Schrödinger equation in (5.4), wherein the set of lowest wells of figure 5.11 is replaced by the single well in figure 5.12 that tracks at large R essentially the saddle value and converges to the double-detachment threshold (Heim and Rau 1996). At small R, this potential coincides with the lowest well, that is, to the lowest eigenvalue λ of the Λ^2 angular momentum operator which dominates at small distances. The lowest states in each of the wells in figure 5.11 are thereby seen as a single sequence in the single potential well of figure 5.12. Although the energy positions are little changed, there is a drastically different picture of the wave functions in the two pictures. In figure 5.11, each state has a radial wave function that is nodeless in R, the higher ones having more nodes in the hyperangles (while, at the same time, all having peak probability density at the saddle point). Instead, all the states in figure 5.12 have no nodes in angles, having traded them for successive nodes in R (see also Bohn 1994 and Bohn and Fano 1996). The two pictures are drastically different for a matrix element such as the one involved in photoabsorption from the ground state. Because of the staggering in R of successive wells in figure 5.11, the wave functions at higher N are very (exponentially) small at small R, which is where the ground state function is concentrated. Therefore, the matrix element and cross-section for photoabsorption decreases exponentially in N on comparing excitation of this set of states. In figure 5.12, on the other hand, the successive states ν have more nodes in R, and the oscillating loops of the wave function at small R have more overlap with the ground state, the photoabsorption cross-section dropping off thereby as a power of ν, just as is indeed observed. Once again, note the one-electron analogy, the description in figure 5.11 being like a sequence $2p, 3d, 4f, \ldots$, all nodeless in r but with more angular nodes, whereas fig. 5.12 is more like $2p, 3p, 4p, \ldots$, that is, with the same angular structure but more radial nodes. Energy considerations alone do not suffice for characterization in a Coulomb problem, given the high degeneracy. One has to examine wave functions as well, and, clearly, the second picture is more in conformity with the excitation of a Rydberg series, now of double excitations (Heim *et al.* 1996; Heim and Rau 1996).

5.5.2 *Threshold escape of the pair*

A sequence such as in figure 5.12 of doubly-excited states connects as $\nu \to \infty$ to the double continuum above the 14.35 eV threshold. The threshold behavior of double escape is, therefore a natural adjunct to the study of highly-excited states (Rau 1971, 1984b). For a one-electron Coulomb problem, the feature that the photoabsorption cross-section to successive n falls off as n^{-3} connects to the finite and constant photoionization cross-section just above threshold (section 2.5 of Fano and Rau 1986). For the two-electron problem, the threshold law was studied by Wannier (1953) long before any doubly-excited states had been observed. He recognized that the correlations between the electrons are at an extreme at threshold given that the electrons are slow, allowing for a long range over which their motion can remain correlated. A joint, pair treatment was therefore essential, and he described one based on classical mechanics. With reference to our discussion and in the language of quantum physics, the final state of

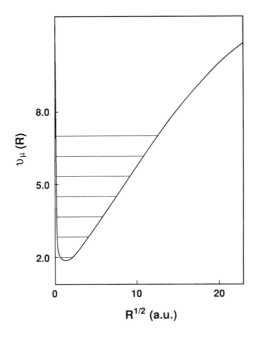

Figure 5.12: Lowest ($\eta = 0$) diabatic potential well for pair states, corresponding to the set in figure 5.11. At small R, it coincides with the lowest of that set while, at larger R, it traces through the loci of avoided crossings to converge to the double-detachment threshold.

double escape has to be handled correctly in terms of the pair's escape to infinity (Rau 1971, 1984c).

It is the final state wave function at large distances that governs the threshold law, and this has to be obtained from (5.4) for $E \geq 0$. The kind of radial and angular correlations that prevail in the so-called "ridge" states described in section 5.5.1, with wave function concentration in the saddle of the potential surface, also pertain to the threshold wave function. Unless the two electrons maintain an equal sharing of the available energy for most of the escape (that is, stay in the vicinity of $\alpha = \pi/4$), double escape will be thwarted, one or the other getting faster at the expense of the other, which will then fall back into a bound state H(N). This instability towards falling away from the saddle into the valleys at $\alpha = 0$ and $\pi/2$ in figure 5.9 acts as a suppression mechanism (Rau 1971). In its absence, the escape of two electrons, described as a product of two Coulomb wave functions, would give a threshold cross-section proportional to E. A "Coulomb-dipole" description (Temkin and Hahn 1974; Temkin 1982), with a Coulomb wave for the inner electron and a dipole wave for the outer electron, would give a similar result (Greene and Rau 1985). The additional suppression in the Wannier pair description raises the exponent from 1; since the saddle potential and departures from it involve Z, so does this exponent. Both Wannier's analysis and a pair hyperspherical treatment of the quantum problem (cf. a review: Rau 1984c) give the value 1.127 for this exponent for $Z = 1$, which applies both to photo double-detachment of H^- (photon energy $= 14.35 + E$) and electron-impact ionization of H (incident electron energy $= 13.6 + E$).

Once again, only the Los Alamos experiment, with its relativistic Doppler amplification of laser photon energies to about 15 eV, has been able so far to study the double detachment. Figure 5.13 shows a cross-section that is indeed compatible with the $E^{1.127}$ threshold law (Donahue *et al.* 1982). Other details of the outgoing electrons, such as the distribution in the mutual angle θ_{12}, or the angular distribution of one electron with respect to the laser polarization, or spin correlations between the electrons, have not been measured. Theoretical predictions exist on all these (Rau 1984c, 1990a; Kato and Watanabe 1995) but the relativistic velocity of the H^- beam makes the electrons emerge in a very forward direction as seen in the laboratory, making their experimental observation difficult. Such measurements will have to await laboratory photodetachment studies once tunable lasers are available in the 15 eV energy range.

The double continuum of H^- can also be studied in electron-impact ionization of hydrogen. Again, unlike similar studies of threshold ionization of He and other rare gases by electrons, no such measurements have been made except for one on the spin dependence of this ionization. The asymmetry parameter for triplet and singlet double-detachment has been studied close to threshold and compared with theoretical predictions (Guo *et al.* 1990; Crowe *et al.* 1990; Friedman *et al.* 1992).

5.6   Collisions with other particles

We have discussed so far the structure of H^- and its interactions with the electromagnetic field. Collisions involving H^- and a second particle have also been studied (see reviews by Risley 1980, 1983; Esaulov 1986). One, already noted at the end of section 5.2.3, is collision with H^+, leading to charge exchange or associative attachment to form H_2^+, processes of interest in stellar atmospheres. Collisions of 1–25 keV H^- ions with Na have also been studied (Allen *et al.* 1988), as well as with noble gases and small molecules over a range of energies from about 100 eV to a few tens of keV (Tuan and Esaulov 1982; Montmagnon *et al.* 1983; Andersen *et al.* 1984). In the case of molecules, charge exchange to shape resonances plays a major role. Below about 1 keV, an adiabatic molecular picture based on the zero-range model of H^- gives an adequate description, whereas at higher energies an impulse approximation for the scattering of the loose

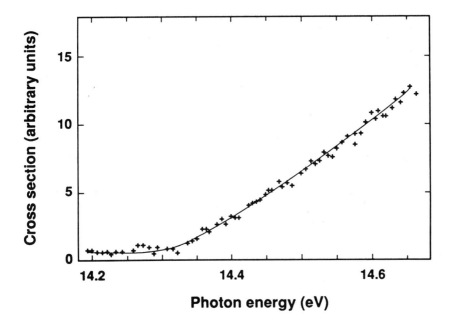

Figure 5.13: Cross-section for photo double-detachment of H^- just above the 14.35 eV threshold (from Donahue *et al.* 1982).

electron suffices. There are also theoretical predictions of novel structures in H^- collisions with atoms (Theodosiou 1991).

Collisions with electrons, also likely in astrophysical contexts, have been studied in the laboratory, particularly with the advent of cooled storage rings for ions. An H^- beam in such a ring is merged with an almost parallel electron beam, so that a low energy process such as detachment, $e + H^- \rightarrow H + e + e$, can be studied. This time, unlike in section 5.5, the double escape is not in a Coulomb but in an asymptotically neutral field. Careful measurements (Andersen *et al.* 1995) show no resonances of states of H^{--}; but, were metastable resonance states to exist (Lieb 1984 and Simon 1974 have proven that no stable bound states exist, and experimental searches (Chang *et al.* 1987) have also been negative), they might be of importance in astrophysics. In analogy with the tightly correlated Wannier state of two electrons, a similar state of three low-energy electrons in a Coulomb potential would be in the saddle point of the potential surface and have the geometrical configuration of an equilateral triangle, with the electrons at the vertices and equidistant from the nucleus at the center. Such a tight correlation, with $r_1 = r_2 = r_3$ and the mutual angles between the $\hat{r}_i$ equal to $2\pi/3$, would only result from a superposition of a very large number of angular harmonics, which might account for why it has escaped notice in theoretical calculations so far (Robicheaux *et al.* 1994).

Finally, the latest experiments from Los Alamos are worth noting, in which the relativistic H^- beam is passed through thin carbon foils (Mohagheghi *et al.* 1991). The time of interaction is less than a femtosecond, the H^- experiencing a pulse of the "matter field." Remarkably, many H^- ions emerge unscathed, although both neutral H and protons H^+ are also observed. The distribution in principal quantum number of the neutrals has been studied, showing a fall-off roughly proportional to n^{-3} for low n ($n = 2$–5) but a much steeper n^{-8} for higher n (10–15). Our understanding of these results and of the interaction of the negative ion with the foil is as yet very incomplete.

5.7   Epilogue

The H^- ion has played an important and central role from the earliest days of quantum physics. As the simplest, and therefore prototype, three-body quantum system with long-range interactions between all pairs of particles, its relevance extends beyond atomic physics to multiparticle problems of chemical transformation and even more general physics. All three regions of energy, the ground state and photoabsorption continuum for visible and near ultraviolet, the low-lying doubly-excited states into the "middle ultraviolet" ($\simeq 10$–13 eV), and the high doubly-excited states and double continuum in the "far ultraviolet" (> 14 eV), exhibit interesting and different effects of electron-electron correlations in this system. This essay has discussed the nature of these correlations and the associated structures, based on the understanding gained from laboratory experiments and theoretical studies over the years. Applications, some central to astrophysics as described by Chandrasekhar over fifty years ago, and the use of H^- as the initial species for acceleration in particle physics accelerators and plasma machines, add further interest to this fascinating species. New techniques that are just emerging, such as far ultraviolet telescopes, detailed coincidence measurements of energy, angular, and spin distributions of two electrons, stored beams of H^- which can be intercepted by other particles over a wide range of collision energies, etc., are likely to give even further insight into the physics of H^- and add to the applications in varied areas of physics and astrophysics.

I had just begun the writing of this essay when my close friend and colleague, the astrophysicist Ganesar Chanmugam, passed away unexpectedly. In our long association, we had many discussions, including some on H^- and its early history. These many memories have been very much in my mind as I wrote and I wish, therefore, to dedicate this first paper after his death to him and to his many contributions.

References

Allen, J. S., Anderson, L. W., Lin, C. C. 1988, *Phys. Rev.*, **A37**, 349.

Andersen, N., Andersen, T., Jepsen, L., Macek, J. 1984, *J. Phys.*, **B17**, 2281.

Andersen, L. H., Mathur, D., Schmidt, H. T., Vejby-Christensen, L. 1995, *Phys. Rev. Lett.*, **74**, 892.

Arrighini, G. P., Guidotti, C., Durante, N. 1987, *Phys. Rev.*, **A35**, 1528.

Avery, J. 1988, *Hyperspherical Harmonics: Applications in Quantum Theory* (Dordrecht: Kluwer).

Bachau, H., Martin, F. 1996, *J. Phys.*, **B29**, 1451.

Bartlett, J. H. 1937, *Phys. Rev.*, **51**, 661.

Bates, D. R. 1990, *Adv. At. Mol. Opt. Phys.*, **27**, 1.

Bethe, H. 1929, *Z. Phys.*, **57**, 815.

Bethe, H., Salpeter, E. E. 1977, *Quantum Mechanics of One- and Two-Electron Atoms* (New York: Plenum).

Bohn, J. L. 1994, *Phys. Rev.*, **A49**, 3761.

Bohn, J. L., Fano, U. 1996, *Phys. Rev.*, **A53**, 4014.

Bray, I., Stelbovics, A. T. 1995, *Adv. At. Mol. Opt. Phys.*, **35**, 209.

Broad, J. T., Reinhardt, W. P. 1976, *Phys. Rev.*, **A14**, 2159.

Bryant, H. C., Dieterle, B. D., Donahue, J., Sharifian, H., Tootoonchi, H., Wolfe, D. M., Gram, P. A. M., Yates-Williams, M. A. 1977, *Phys. Rev. Lett.*, **38**, 228.

Bryant, H. C., Butterfield, K. B., Clark, D. A., Frost, C. A., Donahue, J. B., Gram, P. A. M., Hamm, M. E., Hamm, R. W., Smith, W. W. 1981, in *Atomic Physics 7*, ed. D. Kleppner and F. M. Pipkin (New York: Plenum), 29.

Bryant, H. C., Clark, D. A., Butterfield, K. B., Frost, C. A., Sharifian, H., Tootoonchi, H., Donahue, J. B., Gram, P. A. M., Hamm, M. E., Hamm, R. W., Pratt, J. C., Yates, M. A., Smith, W. W. 1983, *Phys. Rev.*, **A27**, 2889.

Bryant, H. C., Mohagheghi, A., Stewart, J. E., Donahue, J. B., Quick, C. R., Reeder, R. A., Yuan, V., Hummer, C. R., Smith, W. W., Cohen, S., Reinhardt, W. P., Overman, L. 1987, *Phys. Rev. Lett.*, **58**, 2412.

Buckman, S. J., Clark, C. W. 1994, *Rev. Mod. Phys.*, **66**, 539.

Buckman, S. J., Hammond, P., Read, F. H., King, G. C. 1983, *J. Phys.*, **B16**, 4039.

Callaway, J., Rau, A. R. P. 1978, *J. Phys.*, **B11**, L289.

Chandrasekhar, S. 1944, *Astrophys. J.*, **100**, 176.

Chandrasekhar, S. 1945, *Astrophys. J.*, **102**, 223.

Chandrasekhar, S. 1960, *Radiative Transfer* (New York: Dover).

Chandrasekhar, S., Breen, F. H. 1946, *Astrophys. J.*, **104**, 430.

Chandrasekhar, S., Herzberg, G. 1955, *Phys. Rev.*, **98**, 1050.

Chandrasekhar, S., Krogdahl, M. K. 1943, *Astrophys. J.*, **98**, 205

Chandrasekhar, S., Münch, G. 1946, *Astrophys. J.*, **104**, 446.

Chang, K. H., McKeown, R. D., Milner, R. G., Labrenz, J. 1987, *Phys. Rev.*, **A35**, 3949.

Christensen-Dalsgaard, B. L. 1984, *Phys. Rev.*, **A29**, 2242.

Cooper, J. W., Fano, U., Prats, F. 1963, *Phys. Rev. Lett.*, **10**, 518.

Crance, M., Aymar, M. 1985, *J. Phys.*, **B18**, 3529.

Crowe, D. M., Guo, X. Q., Lubell, M. S., Slevin, J., Eminyan, M. 1990, *J. Phys.*, **B23**, L325.

Domke, M., Xue, C., Puschmann, A., Mandel, T., Hudson, E., Shirley, D. A., Kaindl, G., Greene, C. H., Sadeghpour, H. R., Peterson, H. 1991, *Phys. Rev. Lett.*, **66**, 1306.

Domke, M., Schultz, K., Remmers, G., Gutierrez, A., Kaindl, G., Wintgen, D. 1995, *Phys. Rev.*, **A51**, R4309.

Donahue, J. B., Gram, P. A. M., Hynes, M. V., Hamm, R. W., Frost, C. A., Bryant, H. C., Butterfield, K. B., Clark, D. A., Smith, W. W. 1982, *Phys. Rev. Lett.*, **48**, 1538.

Drake, G. W. F. 1973, *Astrophys. J.*, **184**, 145.

Drake, G. W. F. 1974, *Astrophys. J.*, **189**, 161.

Du, M. L., Delos, J. B. 1988, *Phys. Rev.*, **A38**, 5609.

Du, M. L., Fabrikant, I. I., Starace, A. F. 1993, *Phys. Rev.*, **A48**, 2968.

Du, N. Y., Starace, A. F., Bao, M. Q., 1994, *Phys. Rev.*, **A50**, 4365.

Esaulov, V. A. 1986, *Ann. Phys. Fr.*, **11**, 493.

Fano, U. 1981, *Phys. Rev.*, **A24**, 2402.

Fano, U., Rau, A. R. P. 1986, *Atomic Collisions and Spectra* (Orlando: Academic Press).

Feagin, J. M. 1988, in *Fundamental Processes of Atomic Dynamics* ed. J. S. Briggs, H. Kleinpoppen, and H. O. Lutz (New York: Plenum) 275.

Feagin, J. M., Briggs, J. S. 1986, *Phys. Rev. Lett.*, **57**, 984.

Feagin, J. M., Briggs, J. S. 1988, *Phys. Rev.*, **A37**, 4599.

Fock, V. A. 1954, *Izv. Akad. Nauk SSSR, Ser. Fiz.*, **18**, 161 [Engl. transl.: *K. Norsk Vidensk. Selsk. Forh.*, **31**, 138 (1958)].

Friedmann, J. R., Guo, X. Q., Lubell, M. S., Frankel, M. R. 1992, *Phys. Rev.*, **A46**, 652.

Gailitis, M. 1980, *J. Phys.*, **B13**, L479.

Gailitis, M., Damburg, R. 1963, *Sov. Phys.-JETP*, **17**, 1107.

Gavrila, M. 1992, ed., *Atoms in Intense Laser Fields* (Orlando: Academic Press).

Geltman, S. 1990, *Phys. Rev.*, **A42**, 6958.

Geltman, S. 1991, *Phys. Rev.*, **A43**, 4930.

Gram, P. A. M., Pratt, J. C., Yates-Williams, M. A., Bryant, H. C., Donahue, J. B., Sharifian, H., Tootoonchi, H. 1978, *Phys. Rev. Lett.*, **40**, 107.

Greene, C. H. 1987, *Phys. Rev.*, **A36**, 4236.

Greene, C. H., Rau, A. R. P. 1985, *Phys. Rev.*, **A32**, 1352.

Guo, X-Q., Crowe, D. M., Lubell, M. S., Tang. F. C., Vasilakis, A., Slevin, J., Eminyan, M. 1990, *Phys. Rev. Lett.*, **65**, 1857.

Harris, P. G., Bryant, H. C., Mohagheghi, A. H., Reeder, R. A., Tang. C. Y., Donahue, J. B., Quick, C. R. 1990, *Phys. Rev.*, **A42**, 6443.

Heap, S. R., Stecher, T. P. 1974, *Astrophys. J.*, **187**, L27.

Heim. T. A., Rau, A. R. P. 1996 (to be published).

Heim, T. A., Rau, A. R. P., Armen, G. B. 1996 (to be published).

Henrich, L. R. 1944, *Astrophys. J.*, **99**, 59.

Herbst, E., Patterson, T. A., Norcross, D. W., Lineberger, W. C. 1974, *Astrophys. J.*, **191**, L143.

Herrick, D. R. 1983, *Adv. Chem. Phys.*, **52**, 1.

Hill, R. N. 1977a, *Phys. Rev. Lett.*, **38**, 643.

Hill, R. N. 1977b, *J. Math. Phys.*, **18**, 2316.

Ho, Y. K. 1990, *Phys. Rev.*, **A41**, 1492.

Ho, Y. K. 1992, *Phys. Rev.*, **A45**, 148.

Ho, Y. K. 1995, *Phys. Rev.*, **A52**, 375.

Ho, Y. K., Callaway, J., 1984, *J. Phys.*, **B17**, L559.

Ho, Y. K., Callaway, J., 1986, *Phys. Rev.*, **A34**, 130.

Hylleraas, E. A. 1930, *Z. Phys.*, **60**, 624 and **63**, 291.

Kato, D., Watanabe, S. 1995, *Phys. Rev. Lett.*, **74**, 2443.

Koga, T., Matsui, K. 1993, *Z. Phys.*, **D27**, 97.

Koga, T., Morishita, S. 1995, *Z. Phys.*, **D34**, 71.

Koyama, N., Takafuji, A., Matsuzawa, M. 1989, *J. Phys.*, **B22**, 553.

Laughlin, C., Chu, S. I. 1993, *Phys. Rev.*, **A48**, 4654.

Lieb, E. H. 1984, *Phys. Rev. Lett.*, **52**, 315.

Lin, C. D. 1974, *Phys. Rev.*, **A10**, 1986.

Lin, C. D. 1975, *Phys. Rev. Lett.*, **35**, 1150.

Lin, C. D. 1976, *Phys. Rev.*, **A14**, 30.

Lin, C. D. 1983, *Phys. Rev.*, **A28**, 1876.

Lin, C. D. 1986, *Adv. At. Mol. Phys.*, **22**, 77.

Liu, C. R., Gao, B., Starace, A. F. 1992, *Phys. Rev.*, **A46**, 5985.

Lykke, K. R., Murray, K. K., Lineberger, W. C. 1991, *Phys. Rev.*, **A43**, 6104.

Macek, J. H. 1968, *J. Phys.*, **B1**, 831.

Mercouris, T., Nicolaides, C. A. 1993, *Phys. Rev.*, **A48**, 628.

Mohagheghi, A. H., Bryant, H. C., Harris, P. G., Reeder, R. A., Sharifian, H., Tang, C. Y., Tootoonchi, H., Quick, C. R., Cohen, S., Smith, W. W., Stewart, J. E. 1991, *Phys. Rev.*, **A43**, 1345.

Montmagnon, J. L., Esaulov, V., Grouard, J. P., Hall, R. I., Landau, M., Pichou, F., Schermann, C. 1983, *J. Phys.*, **B16**, L143.

Nicolaides, C. A., Komninos, Y., 1987, *Phys. Rev.*, **A35**, 999.

Ohmura, T., Ohmura, H. 1960, *Phys. Rev.*, **118**, 154.

Pathak, A., Kingston, A. E., Berrington, K. A. 1988, *J. Phys.*, **B21**, 2939.

Pathak, A., Burke, P. G., Berrington, K. A. 1989, *J. Phys.*, **B22**, 2759.

Percival, I. C. 1977, *Proc. R. Soc. London*, **A353**, 289.

Praderie, F. 1971, *Astrophys. Lett.*, **9**, 27.

Rau, A. R. P. 1971, *Phys. Rev.*, **A4**, 207.

Rau, A. R. P. 1983, *J. Phys.*, **B16**, L699.

Rau, A. R. P. 1984a, *Comm. At. Mol. Phys.*, **14**, 285.

Rau. A. R. P. 1984b, in *Atomic Physics 9*, ed. R. S. Van Dyck and E. N. Fortson (Singapore: World Scientific), 491.

Rau, A. R. P. 1984c, *Phys. Reports*, **110**, 369.

Rau, A. R. P. 1990a, in *Aspects of Electron-Molecule Scattering and Photoionization*, ed. A. Herzenberg (New York: Am. Inst. of Phys.), 24.

Rau, A. R. P. 1990b, *Rep. Prog. Phys.*, **53**, 181.

Rau, A. R. P. 1992, *Science*, **258**, 1444.

Rau, A. R. P., Fano, U. 1967, *Phys. Rev.*, **162**, 68.

Rau, A. R. P., Molina, Q. 1989, *J. Phys.*, **B22**, 189.

Rau, A. R. P., Wong, H. Y. 1988, *Phys. Rev.*, **A37**, 632 and **38**, 1660.

Read, F. H. 1982, *Aust. J. Phys.*, **35**, 475.

Risley, J. S. 1980, in *Electronic and Atomic Collisions*, Proc. XI ICPEAC (Kyoto: 1979), eds N. Oda and K. Takayanagi, (Amsterdam: North-Holland), 619.

Risley, J. S. 1983, *Comm. At. Mol. Phys.*, **12**, 215.

Robicheaux, F., Wood, R. P., Greene, C. H. 1994, *Phys. Rev.*, **A49**, 1866.

Rost, J. M., Briggs, J. S. 1988, *J. Phys.*, **B21**, L233.

Sadeghpour, H. R. 1991, *Phys. Rev.*, **A43**, 5821.

Sadeghpour, H. R., Greene, C. H. 1990, *Phys. Rev. Lett.*, **65**, 313.

Schwartz, C. 1961, *Phys. Rev.*, **124**, 1468.

Simon, B. 1974, *Math. Ann.*, **207**, 133.

Slonim, V. Z., Greene, C. H. 1991, *Radiat. Eff.*, **122-3**, 679.

Tang, C. Y., Harris, P. G., Mohagheghi, A. H., Bryant, H. C., Quick, C. R., Donahue, J. B., Reeder, R. A., Cohen, S., Smith, W. W., Stewart, J. E. 1989, *Phys. Rev.*, **A39**, 6068.

Tang, C. Y., Bryant, H. C., Harris, P. G., Mohagheghi, A. H., Reeder, R. A., Sharifian, H., Tootoonchi, H., Quick, C. R., Donahue, J. B., Cohen, S., Smith, W. W. 1991, *Phys. Rev. Lett.*, **66**, 3124.

Tang, J., Watanabe, S., Matsuzawa, M. 1992, *Phys. Rev.*, **A46**, 2437.

Temkin, A. 1982, *Phys. Rev. Lett.*, **49**, 365.

Temkin, A., Hahn, Y. 1974, *Phys. Rev.*, **A9**, 708.

Theodosiou, C. E. 1991, *Phys. Rev.*, **A43**, 4032.

Tuan, V. N., Esaulov, V. A. 1982, *J. Phys.*, **B15**, L95.

Vilenkin, N. J. 1968, *Special Functions and the Theory of Group Representations*, Transl. of Math. Monographs, vol. 22 (Providence: Am. Math. Soc.).

Wang, Q., Starace, A. F. 1993, *Phys. Rev.*, **A48**, R1741.

Wannier, G. H. 1953, *Phys. Rev.*, **90**, 817.

Wendoloski, J. J., Reinhardt, W. P. 1978, *Phys. Rev.*, **A17**, 195.

Wigner, E. P. 1948, *Phys. Rev.*, **73**, 1002.

Wildt, R. 1939, *Astrophys. J.*, **89**, 295 and **90**, 611.

Wildt, R. 1941, *Astrophys. J.*, **93**, 47.

Wintgen, D., Delande, D. 1993, *J. Phys.*, **B26**, L399.

Zhang, L., Rau, A. R. P. 1992, *Phys. Rev.*, **A46**, 6933.

Zhou, B., Lin, C. D. 1992, *Phys. Rev. Lett.*, **69**, 3294.

6

S. Chandrasekhar and Magnetohydrodynamics

E. N. Parker

6.1   Introduction

This chapter summarizes the many fundamental contributions of Chandrasekhar to the subject of hydromagnetics or magnetohydrodynamics (MHD), with particular attention to the generation, static equilibrium, and dynamical stability-instability of magnetic fields in various idealized settings with conceptual application to astronomical problems. His interest in MHD seems to have arisen first in connection with the turbulence of electrically conducting fluid in the presence of a magnetic field, sparked by Heisenberg's (1948a,b) formulation of an equation for the energy spectrum function $F(k)$ of statistically isotropic homogeneous hydrodynamic turbulence. From there Chandrasekhar's attention moved to the nature of the magnetic field along the spiral arm of the Galaxy (in joint work with E. Fermi), inferred from the polarization of starlight then recently discovered by Hall (1949) and Hiltner (1949, 1951). The polarization implied a magnetic field along the galactic arm, which played a key role in understanding the confinement of cosmic rays to the Galaxy. The detection and measurement of the longitudinal Zeeman effect in the spectra of several stars (Babcock and Babcock 1955) suggested the next phase of Chandrasekhar's investigations, in which he explored the combined effects of magnetic field, internal motion, and overall rotation on the figure of a star in stationary ($\partial/\partial t = 0$) equilibrium. Chandrasekhar and his students did some of the first work in formulating the quasi-linear field equations for the pressure, fluid velocity, and magnetic field in axisymmetric gravitating bodies. From there his thinking turned to the generation of the magnetic fields of planets and stars by the convective motions of the electrically conducting fluid in their interiors.

Now the outer atmospheres of planets, stars, and galaxies are so tenuous that in most cases the atmospheres do not exert significant forces on the strong external magnetic fields of these objects, so that the external magnetic field is "force-free," i.e., the Lorentz force, given by the divergence $\partial T_{ij}/\partial x_j$ of the Maxwell stress tensor T_{ij}, is negligible. The special properties of these force-free fields provide a particularly elegant mathematical formalism in the axisymmetric case.

Subsequently the challenging problem of laboratory plasmas confined in strong magnetic fields attracted Chandrasekhar's interest, and, with A. N. Kaufman and K. M. Watson, he developed a perturbation solution to the collisionless Boltzmann equation in the strong field limit, applying the solution to the stability of the magnetic pinch.

During this same period of time Chandrasekhar investigated the effect of magnetic fields on the convective instability of an electrically conducting fluid in an adverse temperature gradient. The question of the onset of convection is particularly important in the theory of stellar interiors, because the strong magnetic fields of some stars must surely have effects on the location and strength of the convection and the associated heat transport.

Chandrasekhar's interest in the effect of magnetic fields on the dynamical stability of a convective system led to investigations of the effect on the Rayleigh-Taylor instability and the Kelvin-Helmholtz instability. In the end he organized and compiled his results in a monumental tome entitled *Hydrodynamic and Hydromagnetic Stability* (Chandrasekhar 1961).

It is interesting to note that Chandrasekhar's direct involvement in MHD spanned a period of only twelve years, from about 1949 to 1961 when *Hydrodynamic and Hydromagnetic Stability* was published. Chandrasekhar's research papers are conveniently reprinted in organized form in six volumes (*Selected Papers*, S. Chandrasekhar, University of Chicago Press, 1989) and his work on magnetohydrodynamics is contained in volumes 3 and 4, to which we give reference at appropriate points, indicating the volume number, the paper number, and the page number in sequence within parentheses. The diversity of Chandrasekhar's contributions to MHD can be appreciated only from a detailed catalog of his publications. The present article attempts to provide sufficient perspective and detail within a reasonable span of pages to serve as an outline of the MHD papers in volumes 3 and 4.

6.2   Turbulence

Heisenberg's (1948a,b) heuristic formulation of statistically isotropic homogeneous hydrodynamic turbulence reproduced the basic results of Kolmogoroff (1941a,b) in terms of the energy spectrum function $F(k)$. Heisenberg (1948a,b) constructed a simple nonlinear integral equation for $F(k)$ based on the physical mixing length concept of eddy viscosity. Analytical solution provided the form of $F(k)$ for statistically steady conditions. The result yielded the inertial range $F(k) \sim k^{-5/3}$, extending from the small wave number k_o, at which the motion is driven, down to the viscous cut-off at the large wave number $k_s \sim k_o N_R^{3/4}$ where N_R is the characteristic Reynolds number at the large scale k_o^{-1}. For $k \gg k_s$ Heisenberg's equations provided the tail $F(k) \sim k^{-7}$, whereas in the real world the cutoff beyond k_s is more abrupt. Nonetheless, there was a general feeling of optimism that the old and important problem of hydrodynamic turbulence was at last giving way to solution. The specter of intermittency etc. had not yet come to haunt the theoretical development.

Chandrasekhar was as intrigued as anyone and showed in 1949 (3, 24, 395) how Heisenberg's integral equation for statistically stationary turbulence could be reduced to a linear first order differential equation and one quadrature by a suitable choice of variables. He used $k^3 F(k)$ for the dependent variable and the square of the total vorticity $\int_o^k dk \, k^2 F(k)$ for the independent variable. He went on to treat the more difficult time-dependent free decay of an initial turbulent state.

The next paper (3, 25, 409) picks up on the symmetry of the dynamical terms in the MHD equations to interchanging the velocity v_j and the reduced magnetic field $b_j = B_j/(4\pi\rho)^{1/2}$ in an incompressible fluid. The symmetry is vividly displayed in terms of the Elsasser variables

$$U_i = v_i + b_j, \quad V_j = v_j - b_j,$$

for which the momentum and induction equations take the form

$$\frac{\partial U_j}{\partial t} + V_j \frac{\partial U_j}{\partial x_j} = -\frac{1}{\rho}\frac{\partial P}{\partial x_j} + \frac{1}{2}\nu \, \nabla^2 (U_j + V_j) + \frac{1}{2}\eta\nabla^2 (U_j - V_j),$$

$$\frac{\partial V_j}{\partial t} + U_j \frac{\partial U_j}{\partial x_j} = -\frac{1}{\rho}\frac{\partial P}{\partial x_j} + \frac{1}{2}\nu \, \nabla^2 (U_j + V_j) - \frac{1}{2}\eta\nabla^2 (U_j - V_j).$$

The quantity P represents the total pressure:

$$P = p + \frac{1}{2} \rho b_j b_j.$$

Chandrasekhar proceeded to apply the theory of invariants (Robertson 1940 and (3, 29, 442)), exploited earlier by Batchelor (1950) in connection with hydrodynamic turbulence, to the form of the double and triple correlations of v_j and b_j. He worked out the relations between the scalar functions (coefficients) in the invariant forms for the correlations, obtaining the generalization of the hydrodynamic Von Karman–Howarth equation to MHD, and two additional relations.

The symmetry of the MHD equations in v_j and b_j is complemented by identical forms of the induction equation

$$\frac{\partial \mathbf{b}}{\partial t} = \nabla \times (\mathbf{v} \times \mathbf{b}) + \eta\nabla^2 \mathbf{b}$$

and the vorticity equation in hydrodynamics,

$$\frac{\partial \omega}{\partial t} = \nabla \times (\mathbf{v} \times \omega) + \nu \nabla^2 \omega,$$

where $\omega = \nabla \times \mathbf{v}$. This raises the question of whether there is a useful analogy between $\mathbf{b}$ and ω. Chandrasekhar explored the relation by writing $\mathbf{b} = \nabla \times \mathbf{a}$ in terms of the vector potential $\mathbf{a}$. Then any analogy between ω and $\mathbf{b}$ appears as an analogy between $\mathbf{v}$ and $\mathbf{a}$. Again the application of the theory of invariants provided forms for the double and triple correlations as well as equations relating the various scalar coefficients. But in neither formulation does one obtain enough equations to close the system without introducing additional and arbitrary assumptions. The failure to close is a result of the well known fact that the nonlinear terms in the MHD equations, like the hydrodynamic equations, provide the nth order correlation in terms of the $(n+1)$th order correlations, indicating that there is physics in the equations of $(n+1)$th order that is not contained up to nth order.

Chandrasekhar went on to show that MHD turbulence permits the construction of expressions analogous to the Lotsiansky invariant of hydrodynamic turbulence, based on similar assumptions as to the asymptotic rate of decline of correlations in v_j and in b_j between positions separated by a large distance r.

In stationary MHD turbulence, sustained by the continual addition of kinetic energy at large scales, the scalar coefficients satisfy simpler relations, and a direct analogy to the vorticity correlation $\langle \omega_j (\mathbf{r}) \omega_j (\mathbf{r} + \zeta) \rangle$ is established.

So the double and triple correlations in MHD turbulence are interrelated much as in hydrodynamic turbulence. But, as already noted, the mathematics does not provide a closed system. Some physically motivated form of truncation of the equations is necessary.

We know much more about hydrodynamic and MHD turbulence now, 45 years later, thanks to the work of many theoreticians (cf. Kraichnan 1965), but a comprehensive deductive dynamical theory of turbulence still eludes the best efforts.

6.3   Galactic magnetic field

In the late forties the origin of cosmic rays was a problem of central interest, beginning with their identification as (largely) protons by Schein, Jesse, and Wollan (1941). This led to the question of whether cosmic rays are a local phenomenon confined to the solar system by the dipole magnetic field of the Sun, or a nonlocal phenomenon presumably galactic in extent. Ideas of local confinement were based on a hypothetical highly symmetric solar magnetic dipole with a strength of 50 gauss at the poles of the Sun, suggested by the early work of Hale (1913). A dipole field declining as r^{-3}

extrapolates from 50 gauss at the surface of the Sun to 5×10^{-6} gauss at 1 a.u, with 4×10^7 gauss cm beyond. This is sufficient to deflect a proton of 6 GeV through 180°, from which it follows that a solar dipole field might, in principle, temporarily trap protons of 6 GeV, but not much more. On the other hand, it is observed that cosmic rays arrive at the surface of Earth at the geomagnetic equator, after having penetrated through 10^8 gauss cm in the geomagnetic field. Such particles, with energies in excess of 10 GeV, would not be trapped by the solar magnetic dipole. There was no observed break at 6 GeV in the energy spectrum of the cosmic rays. The cosmic ray intensity varied smoothly with geomagnetic latitude from the equator to the poles. So it appeared that cosmic rays are a galactic phenomenon.

Hiltner's (1949, 1951) studies of the polarization of starlight indicated a magnetic field of at least several microgauss along the local spiral arm. Unfortunately, it was not possible to deduce the strength of the galactic field from the observed polarization without having the precise composition and structure of the spinning interstellar dust grains that provide the polarization. However, Fermi (1949) suggested that cosmic rays are accelerated primarily by bouncing back and forth along the galactic field between reflections from moving magnetic gas clouds. So the structure and dynamics of the galactic magnetic field thrust itself upon the physics community as an important question. In the paper (3, 34, 529) Enrico Fermi and Chandrasekhar addressed the problem of the field strength from the observed dynamical properties of the galactic arm. The polarization studies (Hiltner 1949, 1951) suggested that the rms deflection of the magnetic field is about 0.2 radians. This deflection is presumably dynamical, representing transverse Alfvén waves for which the magnetic amplitude ΔB is related to the transverse amplitude v by $\Delta B = \pm (4\pi\rho)^{1/2} v$ for an interstellar gas density $\rho \sim 2 \times 10^{-24}$gm/cm. An rms isotropic turbulent velocity of 5 km/sec suggested $5/\sqrt{3} \cong 3$ km/sec in the direction transverse to the mean field and to the line of sight, from which they obtained an estimate $B \sim 7 \times 10^{-6}$ gauss.

An alternative value was constructed by estimating the total pressure necessary to support the spiral arm against gravitational collapse. Representing the spiral arm by a circular cylinder of radius R and uniform total mean density ρ_t, they showed that the total pressure on the axis of the cylinder would be $\pi G \rho \rho_t R^2$, where G is the gravitational constant. Then, if half of the total pressure is kinetic, equal to $\rho v^2/3$, and the other half magnetic, equal to $B^2/8\pi$, they obtained 6×10^{-6} gauss, in good agreement with the dynamical result of 7×10^{-6} gauss.

These estimates are about twice the estimates today. The more detailed observational studies since that time suggest that ΔB is more nearly equal to B than to the $0.2B$ assumed in their paper, and the spiral arm is better approximated by a ribbon than a circular cylinder, with a half thickness of 100 pc rather than a radius of 250 pc.

In any case, their effort established the correct order of magnitude, which was more than enough to confine the galactic cosmic rays. The cyclotron radius of a 10 GeV proton moving perpendicular to a magnetic field of 3×10^{-6} gauss is 10^{13} cm, or slightly less than 1 a.u., to be compared with the half thickness of the field, of the order of 100 pc = 3×10^{20} cm = 2×10^7 a.u. To put it differently, a field of 3×10^{-6} gauss in a gaseous galactic disk of half thickness 100 pc represents 10^{15} gauss cm, whereas the deflection of a 10 GeV proton through 180° requires only 3×10^7 gauss cm. From the large-scale dynamical point of view, the cosmic rays, which form a tenuous relativistic gas, exert a pressure of about 0.5×10^{-12} dynes/cm^2, comparable to the pressure of a magnetic field of about $3--4 \times 10^{-6}$ gauss.

Fermi and Chandrasekhar wrote a companion paper (3, 35, 532) on the effect of strong magnetic fields within a star. They used the scalar virial equation

$$\frac{1}{2}\frac{d^2 I}{dt^2} = 2T + 3\left(\gamma - 1\right)U + M + \Omega,$$

where I is the trace of the moment of inertia tensor, T is the total internal kinetic energy, U is the internal thermal energy, and γ is the ratio of specific heats. The total magnetic energy is denoted by M and the total gravitational energy is Ω. The net expansive effect of T, U, and M is obvious here, noting that neither the internal motion T nor the magnetic field M is statistically isotropic. The only negative term on the right-hand side of the scalar virial equation is the gravitational potential energy Ω. They note in passing that equilibrium, obtained by equating the right-hand side to zero, limits the rms field to $1--2 \times 10^8$ gauss within a main sequence star, but no more than a few kilogauss for some expanded giant stars. Then, treating radial pulsations, they pointed out the unbounded increase of the period as the rms field approaches this limiting value. They speculate that such strong magnetic fields may account for the long oscillation periods of some of the giant magnetic stars.

Now, the magnetic fields inside most main sequence stars are nowhere near the theoretical critical values of the order of 10^8 gauss or more. Magnetic buoyancy would bring any such fields to the surface in 10^7 years or less, even if so strong a primordial magnetic field were compressed into the star in the first place. In fact we know from the recent work of Boruta (1996) that the field in the deep interior of the Sun is no more than about 30 gauss. This limit is based on the resistive decay time of 10^{10} years for the basic dipole mode and the fact that there is no fixed dipole in excess of about 5 gauss showing at the surface of the Sun. For in order to confine a dipole field to the interior, it is necessary to superpose higher order radial modes of the dipole. Yet the higher order radial modes decay with periods of 2×10^9 years or less. Since the Sun is about 4.5×10^9 years old, the higher order modes would have decayed away by now, exposing the basic dipole to observation at the surface.

However, Chandrasekhar and Fermi pointed out some newly discovered young giant magnetic stars showing an rms surface field of 2000 gauss and a theoretical maximum internal rms magnetic field of about 3000 gauss. Clearly the magnetic field has a profound influence on the form and behavior of such stars.

They went on in the paper to treat the equilibrium and pulsations of a circular cylinder of self-gravitating fluid of infinite electrical conductivity in which there is a uniform magnetic field parallel to the axis of the cylinder. The effect of the magnetic field is to stabilize the equilibrium, increasing both the minimum wavelength and the growth time of instability.

They showed how the magnetic stresses cause the otherwise spherical form of a star to become oblate in the presence of a dipole magnetic field. Finally, they noted that the criterion for the onset of Jeans' gravitational instability is unaffected by the presence of a uniform magnetic field, because the unstable mode representing motion parallel to the field is unaffected.

The next paper (3, 36, 561), with Nelson Limber, picks up on the pulsation of a star in which a large-scale magnetic field is embedded. They use the time dependent scalar virial equation again, obtaining an approximate expression $\sigma^2 I = -(3\gamma - 4)(\Omega + M)$ for the frequency σ of the oscillations. The moment of inertia I is $4\pi \int dr\, r^4 \rho(r)$. The result shows that σ is real and the star is stable only so long as $M < |\Omega|$, recalling that $\Omega < 0$. The period of oscillation $2\pi/\sigma$ increases without limit as M increases toward $|\Omega|$, as noted in the previous paper with Fermi.

The next several papers involve the MHD equations applied to a star, or other body, with axisymmetry. That is to say, they treat the case in which the magnetic field and fluid motion are independent of azimuth φ measured around some linear axis of the star. The basic nature of the rotating star with a co-aligned magnetic field suggests this idealization as a fruitful starting point for the investigation. The simplification of the MHD equation from 3D to 2D is enormous, although the resulting quasi-linear equations are by no means elementary. So first a word about the general form of the reduction of the dynamical equations in the presence of an ignorable coordinate. The reduction begins by noting that with φ as the ignorable coordinate, the axisymmetric solenoidal vector $\mathbf{B}$ can be decomposed into toroidal and poloidal components, each component represented by a single scalar function of ϖ and z (cylindrical polar coordinates, where $\varpi = (x^2 + y^2)^{1/2}$ represents distance from the z-axis). In terms of the unit vectors $\mathbf{e}_\varpi, \mathbf{e}_\varphi, \mathbf{e}_z$ in the respective coordinate directions, write

$$\mathbf{B}(\varpi, z) = -\mathbf{e}_\varpi \varpi \frac{\partial P}{\partial z} + \mathbf{e}_\varphi \varpi T + \mathbf{e}_z \frac{1}{\varpi} \frac{\partial}{\partial \varpi} \varpi^2 P \tag{6.1}$$

in terms of the scalar function $T(\varpi, z)$ representing the toroidal or azimuthal magnetic field and $P(\varpi, z)$ representing the poloidal or meridional magnetic field. This form guarantees that $\nabla \cdot \mathbf{B} = 0$, thereby reducing the

number of independent functions from three to two. The essential point
for static equilibrium of a gravitating sphere of uniform density is that the
Lorentz force $\partial T_{ij}/\partial x_j$ is balanced by the gradient of the pressure plus the
gravitational potential. Hence the Lorentz force $(\nabla \times \mathbf{B}) \times \mathbf{B}/4\pi$ must have
vanishing curl:

$$\nabla \times (\nabla \times \mathbf{B}) \times \mathbf{B} = 0.$$

In addition, the azimuthal component of the Lorentz force must vanish
because there is no gravitational or pressure force to oppose it.

It is easy to show that

$$
\begin{aligned}
(\nabla \times \mathbf{B}) \times \mathbf{B} = \ & - \left(\Delta_5 P \frac{\partial}{\partial \varpi} \varpi^2 P + T \frac{\partial}{\partial \varpi} \varpi^2 T \right) \mathbf{e}_\varpi \\
& + \left(-\frac{\partial P}{\partial z} \frac{\partial \varpi^2 T}{\partial \varpi} + \frac{\partial T}{\partial z} \frac{\partial}{\partial \varpi} \varpi^2 P \right) \mathbf{e}_\varphi \\
& - \left(\varpi^2 T \frac{\partial T}{\partial z} - \varpi^2 \frac{\partial P}{\partial z} \Delta_5 P \right) \mathbf{e}_z,
\end{aligned}
\tag{6.2}
$$

where Δ_5 represents the axisymmetric Laplacian in five dimensions:

$$\Delta_5 = \frac{1}{\varpi^4} \frac{\partial}{\partial \varpi} \varpi^4 \frac{\partial}{\partial \varpi} + \frac{\partial^2}{\partial z^2}.$$

Setting the φ-component equal to zero requires that

$$\varpi^2 T = F \left(\varpi^2 P \right),\tag{6.3}$$

where F is an arbitrary function of its argument. The consequence of
setting the ϖ-component of the curl of the Lorentz force equal to zero can
be reduced to the Jacobian relation

$$\frac{\partial \left(\Delta_5 P, \varpi^2 P \right)}{\partial \left(\varpi, z \right)} = \varpi \frac{\partial T^2}{\partial z}.\tag{6.4}$$

This equation can be solved using the device that the Jacobian relation

$$\frac{\partial \left(\varpi^2 P, G \left(\varpi^2 P \right) / \varpi^2 \right)}{\partial \left(\varpi, z \right)} = \varpi \frac{\partial T^2}{\partial z}\tag{6.5}$$

defines the function G. This can be seen by writing out the Jacobian, which
reduces to

$$2G \frac{\partial}{\partial z} \left(\varpi^2 P \right) = \frac{\partial}{\partial z} \varpi^4 T^2.$$

If we let $\chi \equiv \varpi^2 P \left(\varpi, z \right)$ and $F = \varpi^2 T$, this can be written

$$2G(\chi) \frac{\partial \chi}{\partial z} = \frac{\partial}{\partial z} F^2,\tag{6.6}$$

and it follows that

$$2G(\chi) = \frac{dF^2}{d\chi}.$$ (6.7)

That is to say, G is determined directly from $F\left(\varpi^2 P\right)$. The purpose of this maneuver is to eliminate $\varpi\partial T^2\partial z$ between equations (6.4) and (6.5), with the result written in the form

$$\frac{\partial\left(\varpi^2 P, \Delta_5 P + G/\varpi^2\right)}{\partial\left(\varpi, z\right) = 0}.$$ (6.8)

The solution is

$$\Delta_5 P + \frac{G\left(\varpi^2 P\right)}{\varpi^2} = \Phi\left(\varpi^2 P\right),$$ (6.9)

where Φ is an arbitrary function of its argument. This field equation for $P\left(\varpi_1 z\right)$ is a quasi-linear elliptic partial differential equation. So the solutions throughout a volume V are uniquely determined by specification of some linear combination of P and ∇P on the surface S enclosing V (Courant and Hilbert 1962).

This simple example serves to illustrate the general method for obtaining the field equations for magnetostatic equilibrium with axisymmetry, which Chandrasekhar pursued at some length. For instance, the paper (3, 37, 565) with K. H. Prendergast works out the field equations and some simple examples of the most general axisymmetric magnetic field that permits static equilibrium of a star of uniform density. The paper (3, 39, 575) extends the formalism to include internal fluid motion. The general conditions deduced in this way prescribe the conditions for hydrostatic equilibrium, the law of isorotation, etc., in a self-gravitating body of uniform density. The paper (3, 45, 632) goes on to apply the general variational principle developed by L. Woltjer to the axisymmetric case, thereby obtaining seven integrals of the field and fluid velocity instead of the four that Woltjer obtained in the general case.

Chandrasekhar makes the important point that the special forms of the field and fluid required by the additional three constraining integrals are not likely to be realized in nature. Three of the seven integrals involve relations between poloidal and toroidal components of the magnetic field and of the fluid velocity. Poloidal and toroidal components tend to have independent physical origins in both the field and fluid motions, and the fluid motion driven by convective forces is not likely to be of such a form as to provide the required relation of the poloidal magnetic field to the toroidal magnetic field and toroidal velocity. Hence one does not expect a convecting magnetic star to achieve a stationary axisymmetric state. This is confirmed by the observed nonuniform distribution of magnetic activity around most stars.

Then the paper (3, 41, 609) formulates the difficult problem of the oscillations of a self-gravitating magnetic star of uniform density in which there is not only an axisymmetric magnetic field but a related fluid velocity **v** everywhere parallel to the magnetic field. Both **v** and **B** are solenoidal in this case, and Chandrasekhar treats the equipartition case $\mathbf{B} = \pm (4\pi\rho)^{1/2} \mathbf{v}$ in which **B** and **v** contribute only to the net pressure, the Maxwell stress (tension) of **B** being precisely offset by the Reynolds stress (compression) of **v**. Expressing both **B** and **v** in terms of their toroidal and poloidal scalar functions, the field equations again reduce to second-order quasi-linear partial differential form. Then a variational principle is used to study the diverse modes of pulsation of a star with toroidal field and flow. Chandrasekhar points out that the method provides only a slow convergence of the result with increasing order of trial functions, but the convergence is sufficient to show that the characteristic pulsations correspond to Alfvén waves propagating around the star. This result expresses the incompressibility of the uniform star, and the Alfvén waves may be thought of as gravity waves since the field tension is canceled to lowest order by the Reynolds stress.

The elegant mathematics of these pioneering papers on axisymmetric static and stationary equilibria of magnetic stars of uniform density ρ sets off in striking manner the much more complicated problem of the gaseous magnetic star with its strong radial stratification, convection, general absence of equilibrium because of magnetic buoyancy and convective overturning, and perpetual nonsteady magnetic activity because of the tendency to form current sheets in all but the simplest field topologies (Parker 1994).

6.4 Generation of magnetic field

A crucial question in the physics of magnetic stars and planets (not to mention interstellar gas clouds, proto-stellar disks, galaxies, and clusters of galaxies) is the origin and maintenance of their magnetic fields. The magnetic fields are continually dissipated through the slight electrical resistivity of the planet or star and, in the case of stars, by the incessant dynamical rapid reconnection of the magnetic field caught up in the internal convection. Indeed, as already noted, even a hypothetical dipole magnetic field anchored in the stable radiative core of the Sun has a characteristic decay time estimated at 10^{10} years. For the planet Earth, the decay time is estimated at $\sim 2 \times 10^4$ years for the dipole mode. The turbulent mixing of magnetic fields in the convective zones of stars may hasten the demise of magnetic fields there, unless, of course, the convection has the special properties sufficient for generating the magnetic field in the first place. In fact one can see from the magnetic cycle of the Sun, and from the comparable magnetic cycles of other stars, that the magnetic field is created and destroyed approximately every decade by the turbulent convection. The creation and destruction can be characterized by a resistive diffusion coefficient η of the order of $10^{11} - -10^{12}$ cm^2/sec. The characteristic decay time is L^2/η for a field of scale L, yielding 10 years for $L \sim 10^{10}$ cm. The diffusion η is conventionally attributed to turbulent mixing of magnetic field, characterized by a mixing length λ and associated eddy velocity $v(\lambda)$, so that $\eta \sim 0.1\lambda v(\lambda)$. However, it is a difficult question how, or whether, the turbulence can perform the assumed mixing and dissipation without producing small-scale magnetic fields vastly greater than the mean macroscopic magnetic field. In fact the mean fields in the convective zone of the Sun are themselves comparable to the equipartition field, so it is not clear why the assumed turbulent mixing and winding of the mean field does not produce small-scale fields of such great intensity as to suppress the convective mixing. The answer seems to be that the effect of the convection is to concentrate the magnetic field into intense filaments or fibrils with the interstices essentially field-free. The individual fibrils are then free to interconnect rapidly across their small diameters.

It is interesting to return to the early days 40 years ago when the problem confronting the theoretician was to establish the limiting conditions for the generation of magnetic fields by the motion of a simply connected body of electrically conducting fluid. Cowling (1934; Bachus and Chandrasekhar 1956) had shown two decades earlier that "when the magnetic field and the fluid motions are symmetric about an axis and the lines of force of the magnetic field as well as the trajectories of the fluid particles are confined to meridional planes, no stationary dynamo can exist." In fact this anti-dynamo theorem was generally understood in a stronger form, that no magnetic field and steady fluid motion with the same topology as with axisymmetry can operate as a self-sustaining dynamo. This stronger conclusion is inferred from the inability to maintain the azimuthal current that necessarily flows through the neutral point (or points) in the poloidal field in the meridional planes. Bachus and Chandrasekhar (1956) proceeded in the paper (3, 38, 570) to provide a formal proof for the ideal axisymmetric case. The proof starts from the fact that the toroidal field necessarily vanishes at the surface of the star or planetary core, whereas the field equation for a stationary field is fully elliptic. Hence the boundary condition at the surface would require that the field vanish throughout. That is to say, no steady state axisymmetric dynamo with uniform conductivity and density exists. Subsequent experience has shown that a variety of dynamo forms exist as soon as one turns to nonsymmetric steady and unsteady flows.

Now if a steady axisymmetric fluid motion cannot sustain a magnetic field, the question arises whether such fluid motion can accelerate or retard the resistive decay of the axisymmetric field. This was taken up by Chandrasekhar in (3, 40, 587), using the established axisymmetric formalism in which each vector quantity is decomposed into its toroidal and poloidal parts. Then the individual modes are found to be expressible in terms of Gegenbaur polynomials $C_n^{3/2}(\cos\theta)$, while the radial dependence is $J_{n+3/2}(kr)/r^{3/2}$ in terms of Bessel functions of half integral order. The intermodal coupling leads to a complicated array of equations. The array is necessarily truncated to effect an asymptotic solution, and Chandrasekhar displayed the convergence of the result as successively more terms were employed. The convergence was clear for weak velocity fields, which is to be expected because such fields are close to the modes of resistive decay in a static fluid. Unfortunately, when the velocity is strong enough to have a substantial effect, the convergence is not so clear. It appeared from the calculations that the decay of both poloidal and toroidal magnetic fields could be slowed by a factor of ten or more by velocity fields that deform the magnetic field so as to decrease the characteristic scale. The calculations also yielded substantially retarded decay in other cases. Chandrasekhar mentions lifetimes increased by factors of 20 or 50. Unfortunately, these interesting cases of prolonged field life are among those exhibiting poor convergence. In fact, a subsequent calculation by G. Bachus (1957) showed formally that no increase in characteristic decay time beyond a factor of four is possible. There is no significant prolonging of the life of a magnetic field without the dynamo effects that generate new field.

Lüst and Schlüter (1954) were the first to emphasize that strong magnetic fields in relatively tenuous gases are of such form that the Lorentz force F_j, i.e., the divergence of the Maxwell stress tensor T_{ij}, is essentially zero:

$$(\nabla \times \mathbf{B}) \times \mathbf{B} = 0.$$

The reason is simply that if the gas is too tenuous to push on the magnetic field, then from Newton's third law it follows that the magnetic field does not push on the gas, $F_j = 0$. The fluid motions, if any, are channeled along the strong field, which acts as a curved conduit of nonuniform cross-section in the general case. So the force-free condition is a restriction on the field, requiring

$$\nabla \times \mathbf{B} = \alpha \mathbf{B} \tag{6.10}$$

in general, where α is a scalar function of position, constant along each field line ($\mathbf{B} \cdot \nabla\alpha = 0$) but varying arbitrarily from one line to the next. The fluid moves freely along the field without significant effect on the field.

Now theory shows that the fields cannot be force-free everywhere throughout a body of gas. The simple scalar virial equation, noted above, shows that the overall effect of the magnetic field is measured by the total magnetic energy, which is positive definite. So the magnetic field engenders expansion and the field can be in static equilibrium only if held firmly in the grip of the negative gravitational potential of a star or other gravitating body. So the Lorentz force may vanish to give a force-free field in the region outside the gravitating body, but it must be remembered that the Lorentz force cannot vanish everywhere inside the body.

Lüst and Schlüter treated the special case of an axisymmetric force-free field with $\alpha = constant$ to illustrate the properties of the force-free field. In the paper (3, 42, 618) Chandrasekhar wrote down the general solution for that illustrative case, in terms of Gegenbauer polynomials and Bessel functions of half integral order. He went on to treat the boundary conditions at the surface of a spherical shell adjoining another shell in which the constant value of α is different. The calculations show the interesting result that the energies of the poloidal and toroidal field components are equal.

The next paper (3, 43, 623), with P. C. Kendall, extends the calculations to the resistive decay of the force-free poloidal and toroidal modes in the presence of uniform resistivity, showing that the decay preserves the force-free form of the field, a general point first made by S. Lundquist (1952). Thus no fluid motions are created as a consequence of the resistive decay.

The paper (3, 44, 627), by Chandrasekhar and L. Woltjer, takes up the question of the field configuration with the maximum magnetic energy, i.e., the maximum mean square magnetic field, for a fixed mean square current density. They pointed out that there can be no minimum mean square field for a given mean square current density, because the mean square current density can be made arbitrarily large without affecting the mean square field by the simple procedure of introducing many steep gradients or shears in the magnetic field. The variational problem is easily formulated, maintaining the volume integral of $(\nabla \times \mathbf{B})^2$ constant while the integral of $(\mathbf{B})^2$ is an extremum. With Lagrangian multiplier α^2 the final result is the elliptic equation

$$\nabla^2\mathbf{B} + \alpha^2\mathbf{B} = 0$$

encompassing the force-free fields with constant α, as well as other solutions. However, it should be noted that the conditions for static equilibrium are not incorporated into the derivation. So the only equilibrium field for which the magnetic energy is maximum for a given mean square current density is the force-free field with constant α.

Note again that the magnetic field cannot be force-free everywhere. For the field must be confined by inward forces if it is not to expand to infinity. In star-like structures one would expect to find either that the field is held in the grip of the central core, or that it is otherwise confined by some hypothetical enclosing boundary outside the force-free regions, or both.

6.5   Collisionless plasmas

Laboratory plasma confinement is achieved by surrounding a volume of plasma with a strong magnetic field. The scales are not astronomical and indeed the characteristic scale perpendicular to the confining magnetic field may be not many times larger than the cyclotron radius of the ionic component of the plasma. The convenient idealizations of MHD, treating the very large-scale behavior as the dynamics of an electrically conducting isotropic fluid, become poor approximations. The thermal velocity distribution is generally not isotropic for a variety of reasons, e.g., the free motion of charged particles along the field as distinct from the cyclotron motion perpendicular to the field. The free particle motion along the field is reflected from regions of strong field by the invariant diamagnetic moment $mw_\perp^2/2B$ of the particle with mass m and velocity $w_\perp$ perpendicular to B. The cyclotron motion of the ions and electrons around $\mathbf{B}$ provides a drift of the guiding center (the instantaneous center of the cyclotron circular motion) perpendicular to $\mathbf{B}$ as a consequence of the curvature of the field lines (the curvature drift) and as a consequence of the variation of the field intensity in the direction perpendicular to B (the gradient drift). In view of the free interpenetration of particles from different regions along the field, where the curvature and field gradients as well as the thermal velocities may be quite different, the general dynamics of the confined plasma presents a daunting problem.

Chandrasekhar, with A. N. Kaufman and K. M. Watson, took on the problem in the two papers (4, 1, 3) and (4, 2, 39), neglecting Coulomb interactions between particles (the collisionless plasma) and working in the strong field limit so that the plasma introduces only a small perturbation of the magnetic field. Thus, the calculation omits thermalization of the ions and electrons, and is a valid representation of the plasma dynamics over periods that are short compared to the thermalization or collision time. Even so, the formal calculation is massive, starting with the collisionless Boltzmann equation (the Vlasov equation)

$$\frac{\partial f}{\partial t} + v_j \frac{\partial f}{\partial x_j} + \left[g_i + \frac{e}{m} \left(E_j + \frac{1}{c} \epsilon_{ijk} v_j B_k \right) \right] \frac{\partial f}{\partial v_j} = 0 \qquad (6.11)$$

for the velocity distribution functions $f(x_j, v_k, t)$ of the individual ions and electrons in the presence of a gravitational acceleration g_j and the electric and magnetic fields E_i and B_j, respectively. They wrote $v_j = V_j + w_i$ where

$$V = \int d^3 v_k f v_i \qquad (6.12)$$

is the local mean velocity and w_j is the thermal velocity. The collisionless Boltzmann equation was then written in a variety of forms, e.g., equation (22) of (4, 1, 3), which was cast in the form[1]

$$\frac{\partial}{\partial t} \left(\sum_{+-} mNV_i + \frac{B^2}{4\pi c^2} \Psi_i \right) = \frac{\partial}{\partial x_i} \left(T_{ij} + P_{ij} - \sum_{+-} mNV_iV_j \right) + \rho g_i.$$
$$(6.13)$$

---

[1] A factor $1/c$ is missing from the term $\epsilon_{ijk} E_j B_k$ in equation (22).

Here $\Psi_i = c\epsilon_{ijk}E_jB_k/B^2$ is the so called electric drift velocity, T_{ij} is the Maxwell stress tensor

$$T_{ij} = -\delta_{ij}\frac{E^2 + B^2}{8\pi} + \frac{E_iE_j + B_iB_j}{4\pi},$$

reducing to

$$T_{ij} \cong -\delta_{ij}\frac{B^2}{8\pi} + \frac{B_iB_j}{4\pi} \tag{6.14}$$

for $v_i \ll c$. P_{ij} is the total pressure tensor (ions and electrons)

$$P_{ij} \equiv \sum_{+-} m \int d^3w_k f w_i w_j, \tag{6.15}$$

and the summation is over both electrons and ions. The electric drift velocity Ψ_j derives from the Poynting flux $c\epsilon_{ijk}E_jB_k/4\pi$. Its contribution to the momentum density on the left-hand side is of the order of the magnetic energy density divided by the rest energy density of the particles. This is not small in the limit of tenuous plasma, of course, but it is generally small when the gas is dense enough that the Alfvén speed is small compared to c.

The time dependent Boltzmann equation is treated for small perturbations about a stationary state. The electromagnetic field perturbations are expressed in terms of the Lagrangian displacement of the artificial velocity U_j defined by the equation of motion

$$\frac{\partial U_i}{\partial t} = \frac{e}{m}\left(E_i' + \frac{1}{c}\epsilon_{ijk}U_jB_k^0\right), \tag{6.16}$$

where B_k^0 represents the stationary field and the prime denotes the perturbation, with

$$B_j = B_j^0 + B_j', \quad E_i = E_i^0 + E_i'. \tag{6.17}$$

The calculation proceeds from there to work out the general conditions for the stationary fields B_j^0, E_j^0, treating the particle motion essentially in the guiding center approximation, as well as developing the macroscopic boundary conditions at a discontinuity. The second paper (4, 2, 39) works out the pressure drift, which is a combination of the gradient drift of the individual particles and the net local particle cyclotron motion in the presence of a nonvanishing cyclotron radius and a plasma pressure gradient. The paper goes on to describe the general plasma conditions in a variety of special conditions.

The final paper (4, 3, 64), with A. N. Kaufman and K. M. Watson, treats the stability of the laboratory magnetic pinch. Rosenbluth (1957) had previously treated the problem using the particle orbits in the guiding center approximation in place of the Boltzmann equation. The more detailed study from the Boltzmann equation gives a slightly different criterion for marginal stability, but the principal results for stabilizing the pinch are confirmed. The calculation and the ultimate stability criteria for the various modes, both immensely complex, are best studied from the original paper. No attempt to summarize the results can be made without a detailed description of the formalism.

The invariants of the guiding center motion of a charged particle in a strong magnetic field are described in a subsequent paper (4, 4, 85) by Chandrasekhar, which the reader may find useful to have in mind when studying the three papers just mentioned. The strongest invariant is the diamagnetic moment μ of the cyclotron motion of the particle (ion or electron) around the field. If $w_\perp$ denotes the particle velocity perpendicular to the field, we have $\mu = Mw_\perp^2/2B$. The invariance of μ can be violated only by changes in the field over scales comparable to or smaller than the cyclotron radius $Mw_\perp c/eB$ or over times less than the cyclotron period $2\pi\, Mc/eB$.

The longitudinal invariant is $\oint ds \cdot \mathbf{w}_\parallel$, where $\mathbf{w}_\parallel$ is the particle velocity parallel to the magnetic field. The integration over length ds along the field is carried out from one mirror point (where the particle is reflected from a region of increasing B) to the other. The invariance of this quantity is preserved for changes in the field that take place over characteristic times that are large compared to the bounce time of the particle between mirror points. The concept and validity of the invariants of various orders are discussed at length in this paper.

The reader who is not already familiar with the guiding center orbit theory of particle motions and with the associated invariants may find the small book on plasma physics (Chandrasekhar and Trehan 1960) a useful place to begin. The book goes on to give a simplified and lucid treatment of the stability of the pinch before taking up plasma oscillations and transport phenomena in the collisionless plasma.

6.6   Magnetic fields and convective instability

Fluids are subject to a variety of dynamical instabilities. A static fluid undergoes convective overturning if heated from below or cooled from above. In general an adverse vertical density stratification may be caused by a temperature or compositional gradient, producing a Rayleigh-Taylor instability and the associated overturning of the fluid. The presence of a directed radiation field and a spatially varying opacity may induce unstable temperature and density distributions. The relative motion of two contiguous volumes of fluid produces a Kelvin-Helmholtz instability at the interface. These instabilities all arise from the interplay of fluid pressure, gravitational acceleration, and Reynolds stress $R_{ij} = -\rho v_i v_j$. The Reynolds stress is a compressive force ρv^2 in the direction of v_i, causing buckling of the stream lines to produce the Kelvin-Helmholtz instability.

The presence of a magnetic field in an electrically conducting fluid adds the Maxwell stress, represented by T_{ij}, described by equation (6.14). In particular, the magnetic field introduces an isotropic pressure $B^2/8\pi$ (B in gauss) and a tension $B^2/4\pi$ along the field. The tension in the magnetic field tends to stabilize waves with phase along the magnetic field, as distinct from the Reynolds stress compression which destabilizes waves with phase along the velocity field. The magnetic pressure tends to expand a compressible (gaseous) fluid, providing buoyancy in the presence of a gravitational field. The buoyancy of the magnetic field contributes a form of the Rayleigh-Taylor instability. Then, of course, in a rotating body the fluid velocity gives rise to a Coriolis force $2\mathbf{v} \times \mathbf{\Omega}$, whereas the magnetic field $\mathbf{B}$ produces no comparable effect. The tension in the field strives merely to make everything rotate with the same angular velocity along each field line.

It is evident from these brief remarks that the subject of hydrodynamic stability and instability takes on new dimensions in the presence of electrical conductivity and a magnetic field. Clearly a methodical recalculation of the classical hydrodynamic instabilities was in order, with the expectation of new instabilities as well as the suppression of familiar hydrodynamic instabilities by the tension in the field. Chandrasekhar's lifelong interest in stars led to a concern with thermal convection, so the general magnetohydrodynamical theory of convection was an obvious challenge. The customary starting point is a fluid of uniform density except for a small thermal expansion coefficient which provides the buoyant forces that drive the convection. The slight thermal density change has no sensible effect on the inertia of the fluid (the Boussinesq approximation). The classical Bernard problem of convection was studied a century earlier by Rayleigh, and by many others since. The reader is referred to Chandrasekhar's (1961) comprehensive monograph for a detailed discussion of the historical development of the theory of thermal convection. The application of convection to stellar structure immediately introduces the theoretical problem of convection in a rotating system. This suggests convection in the presence of both rotation and a magnetic field with no particular special relative orientation of the gravitational acceleration g, the angular velocity $\mathbf{\Omega}$, and the magnetic field $\mathbf{B}$.

To begin with the simpler cases, then, Chandrasekhar (1953; Chandrasekhar and Elbert 1955) investigated the effect of rotation on the dynamics of thermal convection. The results are concisely summarized in Chandrasekhar's Rumford Medal Lecture in 1957 (4, 8, 163), where he begins by noting that the rotation strongly constrains the fluid motion. The effect is stated by the Taylor-Proudman theorem that *all slow motions* (for which the nonlinear terms can be neglected) *in a rotating inviscid fluid are necessarily two dimensional*, being invariant in the direction of the uniform angular velocity of the body of fluid. It follows that an inviscid fluid is stable against convective overturning by an adverse temperature gradient in the direction of the angular velocity, no matter how strong the temperature gradient. The introduction of viscosity, on the other hand, vitiates the Taylor-Proudman theorem and provides convective instability in a suitably strong temperature gradient. In a rotating system, the convective instability may appear as an overstability, in which the motion is oscillatory (as in a stable system) but the amplitude of the oscillations grows exponentially with time. The system is overstable at small Prandtl numbers ν/κ (where ν is the kinematic viscosity and κ is the thermometric conductivity) and unstable at large Prandtl numbers.

In the Rumford Lecture, Chandrasekhar (1952, 1954a) pointed out that the introduction of a magnetic field parallel to gravity and angular velocity tends to stabilize the electrically conducting fluid, for the simple reason that the vertical magnetic field inhibits any variation of the horizontal fluid velocity with height, pushing the system back toward the Taylor-Proudman condition. If we supposed that the layer of fluid is capped above and below by rigid infinitely conducting boundaries, instead of free boundaries, the field is line tied at the boundaries so that the field inhibits all motion, of course. For instance, in applications to sunspots the field lines are largely free to be moved about at the upper end of the sunspot field (at the visible surface), being tied only at the distant opposite end of the bipolar field configuration. The field is tied into the convective motions at the bottom end, where the lines are subject to some unknown pattern of circulation. With such strong magnetic fields, the convective motions are largely constrained to vertical oscillations along the field. The general effect is to inhibit convective heat transport, thereby producing a cool region at the visible surface. One can imagine the endless variety of circumstances that arise in the presence of the three independent vectors g, Ω, and $\mathbf{B}$, together with the Prandtl number, Rayleigh number, and magnetic Reynolds number (cf. Chandrasekhar 1954b, 1956). Chandrasekhar pointed out the somewhat different and conflicting roles of Ω and $\mathbf{B}$, with the possible overstability from both Ω and $\mathbf{B}$ in certain parameter ranges, and instability in other ranges. The combination (discussed at some length in chapter 5 of Chandrasekhar 1961) provides a number of distinct circumstances. In the paper (4, 9, 192) the overstability is addressed from the energy or thermodynamic point of view. The purely mathematical aspects of the theory of hydrodynamic and hydromagnetic (MHD) instability are treated in the paper (4, 11, 207) on characteristic value problems and the paper (4, 12, 221) on adjoint differential systems and variational principles. There is extensive discussion to be found at several places in the monograph (Chandrasekhar 1961).

The foregoing labors were all theoretical, of course, involving a variety of mathematical techniques and enormous algebraic undertakings. It is interesting to note, then, that at the same time an experimental effort was launched at the University of Chicago to test the theoretical predictions. The project was initiated under the auspices of Professor S. K. Allison, who was Director of the Institute for Nuclear Studies (now the Enrico Fermi Institute). Professor D. Fultz carried through a number of experiments of convection in rotating systems—the rotating dishpan experiments (Fultz and Nakagawa 1955; Nakagawa and Frenzen 1955). Dr. Y. Nakagawa carried on the effort with the addition of uniform magnetic fields, up to about 8000 gauss, between the pole pieces of a 36-inch cyclotron magnet. The cyclotron had been decommissioned some time earlier and the magnetic yoke and pole pieces were reconditioned and put to use again. Nakagawa used mercury in depths of a few centimeters, achieving magnetic Reynolds numbers rather less than one. Chandrasekhar worked closely with the experimenters and communicated several of the experimental papers for publication in the Proceedings of the Royal Society. Nakagawa (1957) showed the close agreement of theory and experiment in the presence of a magnetic field **B**. A year later he exhibited results of combined Ω and **B** (Nakagawa 1959), generally confirming the validity of the theoretical predictions.

6.7   Magnetic fields and dynamical instability

Chandrasekhar's contributions to the effect of magnetic fields on the dynamical instability of Couette flows, the Rayleigh-Taylor instability in adverse density gradients, and the Kelvin-Helmholtz instability between fluids with relative tangential velocity are summarized in the aforementioned monograph (Chandrasekhar 1961). The stationary flow between concentric cylinders in relative rotation is an example of Couette flow. The fluid velocity is entirely azimuthal and a function only of distance ϖ from the axis of rotation. Under steady conditions the torque (in the axial direction) transmitted by the viscosity is independent of ϖ, from which it is readily shown that $v(\varpi) \sim 1/\varpi$ in the presence of a uniform viscosity. Rayleigh pointed out a century ago that Couette flow is stable if the angular momentum density $\rho\varpi v(\varpi)$ increases outward and unstable if it decreases outward. We note that for uniform density and viscosity the angular momentum density is independent of radius, providing neutral stability. On the other hand, if viscosity is neglected, then any variation of v with radius is possible, providing both stable and unstable Couette flow. The dynamical effects can be strikingly different in different cases, and the interested reader is referred to Chandrasekhar's monograph. Chapter 9 of the monograph takes up the stability for a conducting fluid with a uniform magnetic field parallel to the axis of rotation, an azimuthal magnetic field (parallel to the azimuthal velocity v), and a combination of axial and azimuthal fields, with and without viscosity. The magnetic tension tends to stabilize the system, of course, and the detailed effects are different in each special case.

The Rayleigh-Taylor instability of superposed fluids arises when the upper fluid is denser, so that gravitational potential energy is released by interchanging or overturning fluid. The effects of vertical magnetic fields and of horizontal magnetic fields are treated in chapter 10, with the tension in the magnetic field inhibiting the onset of instability. Short wavelengths are most strongly inhibited by a vertical magnetic field, so that the growth rate does not increase without bound with increasing wave number, as it does in the inviscid nonconducting case. The inhibition declines to zero in the limit of long wavelengths, of course. The stabilizing effect of a horizontal magnetic field is equivalent to the effect of surface tension.

Finally, the influence of a magnetic field on the Kelvin-Helmholtz instability is treated in chapter 11, with similar results. The tension in the field tends to stabilize any waves with phase extending along the field, with the consequence that the velocity difference between the two relatively moving semi-infinite regions of fluid must exceed the Alfvén speed to produce instability. For fluids of different densities ρ_1 and ρ_2, the final result is slightly more complicated because there is no single Alfvén speed, but the principle is the same, that the system is stable when the tension in the field exceeds the Reynolds compressive stress. The magnetic field perpendicular to the direction of flow has no effect on the unstable waves with wave vector parallel to the flow.

6.8   Concluding remarks

In conclusion one can only remark on the vast and various contributions that Chandrasekhar has made to magnetohydrodynamics. The present article is only the briefest summary of the many different problems elucidated by Chandrasekhar's theoretical studies. The importance of his contributions can be comprehended at the most primitive level by noting that his monograph on *Hydrodynamic and Hydromagnetic Stability* (Chandrasekhar 1961) has sold n copies with $\ln n \sim 11$. The monograph has been reprinted now by Dover Publications of New York. It must be appreciated that the monograph covers only a modest part of Chandrasekhar's contributions to hydromagnetics or MHD. The publication by Dover is not without practical significance to the scientific community, and it was not without personal significance to Chandrasekhar, who recognized the important scientific role of Dover Publications in reprinting landmark books after they have passed out of print on the regular market. This point is best made by relating an experience of some 35 years ago. I was a junior faculty member of the Physics Department at the University of Chicago. One morning, walking to my office, I met Chandrasekhar coming the other way. He was in good spirits, and as we met he said, "Well, Parker, I have been immortalized." To my puzzled look he added "Dover has decided to publish my *Radiative*

Transfer." And, as we all know, Dover went on to publish several of his monographs, which make excellent textbooks to this day.

References

Babcock, H. W., Babcock, H. D. 1955, *Astrophys. J.*, **121**, 349.

Bachus, G. 1957, *Astrophys. J.*, **125**, 500.

Bachus, G. E., Chandrasekhar, S. 1956, *Proc. Nat. Acad. SW*, **42**, 105.

Batchelor, G. K. 1950, *Proc. R. Soc. London*, **A201**, 405.

Boruta, N., 1996, *Astrophys. J.* (in press).

Chandrasekhar, S. 1952, *Phil. Mag.*, **7, 43**, 501.

Chandrasekhar, S. 1953, *Proc. R. Soc. London*, **A217**, 306.

Chandrasekhar, S. 1954a, *Phil. Mag.*, **45**, 1177.

Chandrasekhar, S. 1954b, *Proc. R. Soc. London*, **A225**, 173.

Chandrasekhar, S. 1956, *Proc. R. Soc. London*, **A237**, 476.

Chandrasekhar, S. 1961, *Hydrodynamic and Hydromagnetic Stability* (Oxford: Oxford University Press).

Chandrasekhar, S., Elbert, D. F., 1955, *Proc. R. Soc. London*, **A231**, 198.

Chandrasekhar, S., Trehan, S. K., 1960, *Plasma Physics* (Chicago: University of Chicago Press).

Courant, R., Hilbert, D. 1962, *The methods of mathematical physics* (New York: Wiley), vol. 2, 154.

Cowling, T. G. 1934, *Mon. Not. R. Astron. Soc.*, **94**, 39.

Fermi, E. 1949, *Phys. Rev.*, **75**, 1169.

Fultz, D., Nakagawa, Y., 1955, *Proc. R. Soc. London*, **A231**, 211.

Hale, G. E. 1913, *Astrophys. J.*, **38**, 27.

Hall, J. S. 1949, *Science*, **109**, 166.

Heisenberg, W. 1948a, *Zeit. Phys.*, **124**, 628.

Heisenberg, W. 1948b, *Proc. R. Soc. London*, **A195**, 402.

Hiltner, W. A. 1949, *Astrophys. J.*, **109**, 471.

Hiltner, W. A. 1951, *Astrophys. J.*, **114**, 241.

Kolmogoroff, A. N. 1941a, *C. R. Acad. Sci. URSS*, **30**, 301.

Kolmogoroff, A. N. 1941b, *C. R. Acad. Sci. URSS*, **32**, 16.

Kraichnan, R. H. 1965, *Phys. Fluids*, **8**, 1965.

Lundquist, S. 1952, *Arkiv. Fysik*, **2** (35).

Lüst, R., Schlüter, A. 1954, *Z. Astrophys.*, **34**, 263.

Nakagawa, Y. 1957, *Proc. R. Soc. London*, **A140**, 108.

Nakagawa, Y. 1959, *Proc. R. Soc. London*, **A249**, 138.

Nakagawa, Y., Frenzen, P. 1955, *Tellus*, **7**, 1.

Parker, E. N. 1994, *Spontaneous current sheets in magnetic fields* (Oxford: Oxford University Press).

Robertson, H. P. 1940, *Proc. Camb. Phil. Soc.*, **36**, 209.

Rosenbluth, M. 1957, *Stability of the pinch* (Los Alamos Scientific Laboratory Report No. 2030).

Schein, M., Jesse, W. P., Wollan, E. O. 1941, *Phys. Rev*, **59**, 615 (letter).

7

The Virial Method and the Classical Ellipsoids

Norman R. Lebovitz

7.1 Introduction

Chandra completed his book *Hydrodynamic and Hydromagnetic Stability (HHS)* in 1960 and, as was his custom, turned his attention to a new area of research. He began to study general relativity at this time, and it appeared that this would be his exclusive new direction. However, as the result of a pair of accidents, Chandra in fact devoted much of the period from 1960 through 1968 to the virial method and to an analysis of the figures of the classical ellipsoids and their stability. This subject and the general theory of relativity competed for his attention during these years. It was only after completion of his book *Ellipsoidal Figures of Equilibrium (EFE)* (Chandrasekhar 1969) in 1968 that he felt able to devote himself primarily to the subject of relativity, which then was the principal occupation of the remainder of his research career. His enthusiasm for the development of the classical ellipsoids waxed and waned during this period, and he wrote that parts of it were performed "under protest," his sense of responsibility to the subject taking precedence over his inclination to enter more fully into the study of relativity.

The virial theorem, in scalar form, has a history in astronomy (cf. Ambartsumian 1958). In the theory of stellar pulsations, it was employed by Ledoux (1945) to obtain an approximate expression for the lowest mode of radial pulsation and the effect on that mode of a slow rotation. Tensor forms of the virial theorem had been employed by Rayleigh (1903) and by Parker (1957) in special contexts, and Chandra had long had in the back of his mind the notion that one could use this form of the theorem to obtain useful, approximate information about figures seriously distorted from the spherical by rotation or magnetic fields. He had included the basic equations in his preparation of *HHS* with this in mind. I was at this time one of his research students and it was therefore natural that he suggest to me, as part of my dissertation, the development of this method. The application he proposed was the problem of the oscillations and stability of the Maclaurin spheroids.

This would represent a test case: the frequencies of the Maclaurin spheroids were known, and the virial equations, along with a linear *ansatz* for the Lagrangian displacement as in Ledoux's problem, would lead to approximate frequencies which could then be compared with the exact values. The ansatz would be needed because the virial method is a moment method which would require some kind of approximate closure procedure, such as that provided by the ansatz. What neither of us anticipated was the discovery, in winter 1960–61, that the virial equations, in the context of these incompressible figures, form a closed system and therefore give the *exact* frequencies, without the need for an ansatz (Lebovitz 1961). This was the first accident diverting Chandra's attention from relativity to the virial theorem and the classical ellipsoids: although he had hoped the virial method would be powerful, he now realized that it was more powerful than he had expected it to be. It presented an elementary alternative to the analysis via expansions in ellipsoidal harmonics, which was the method employed in the lengthy and arduous analyses of the Jacobi ellipsoids carried out in the latter part of the nineteenth century and the early part of the twentieth century by Poincaré, Darwin, Lyapunov, and Jeans. This new method should allow one to simplify and extend these analyses. This powerful technique needed to be developed more fully. And this he began to do in earnest, as described below, developing the machinery needed to apply the virial method to rotating, self-gravitating masses, and applying this machinery to study the linearized stability of both the Maclaurin and the Jacobi sequences (to verify and extend the classical analyses) and of compressible, rotating masses as well.

This program had come to a stage of apparent completion in the spring of 1964, and a summarizing paper (Chandrasekhar and Lebovitz 1964) had even been written. Chandra was invited at this time to speak at the Courant Institute in New York City when, browsing during an hour of leisure in Stechert's bookstore, he chanced upon a copy of Bassett's *Hydrodynamics* (Bassett 1888), and purchased it. This was the second of the pair of accidents. Bassett's book contains an account of the Riemann ellipsoids, discovered and discussed by Dirichlet, Dedekind, and Riemann in the period 1857–1861—long before the work of Poincaré and others on the Jacobi ellipsoids, but barely alluded to in their work, and unknown to Chandra until he looked into Bassett's book. The Riemann ellipsoids represent a substantially more general family of solutions of the equations of the fluid dynamics of self-gravitating figures than the Maclaurin and

Jacobi families, and their properties had been much less fully explored. On the one hand, one's understanding of the possible figures of self-gravitating masses and their stability was evidently much narrower than it could be—and should be—and, on the other hand, he now had sufficient technique to bring the understanding of these more general ellipsoids to the point of development that had been achieved for the Jacobi and Maclaurin families. He resolved to bring this beautiful but neglected theory to a fuller stage of completion.

In 1963, Chandra had given the Silliman lectures at Yale University on the subject of "The Rotation of Astronomical Bodies." These lectures were to be written up in book form. Chandra put off doing this during the period when the study of the Riemann ellipsoids was taking place and subsequently used this opportunity to expand the lectures into his book *EFE*, which encompassed his own research and that of his students and collaborators over the period in question. On its completion in 1968, his formal association with the subject ended.

Chandra was deeply interested in scientific and intellectual history and in the motivations of successful scientists, scholars and artists. He admired the funeral essays given by serious scientists of the past on the subject of colleagues who had recently died. Those that he admired most were not eulogies but rather analyses of the contributions of the scientific personality who had recently died by someone able to place those contributions in a general scientific perspective. Indeed, Chandra himself was impatient with fulsome but vague praise of his own work, and preferred constructive criticism based on an understanding of the subject. The practice of funeral essays, a feature of a more leisurely era, has lapsed, and it is in any case difficult to imagine any one individual able to place Chandra's diverse contributions into perspective. The current volume, however, may indeed serve the kind of purpose Chandra would have admired and respected.

Over the years, from time to time, he wrote a chronology of his research efforts during a certain space of time, together with remarks on the scientific and personal background for this period of his research. A number of these he copied and sent to me. I have benefited in writing the current article from his own observations for the period from 1960 to 1968.

Sections 7.2 and 7.3 below present background to the subject matter: a brief history of the classical ellipsoids and an explanation of the virial method with its advantages and disadvantages. Section 7.4 provides a description of some of Chandra's contributions during this period (roughly 1960–1968), and, finally, section 7.5 provides a retrospective view including a sampling of subsequent developments. I have not attempted to include a systematic bibliography, since this can be found in *EFE*.

7.2 A History of the ellipsoidal figures

Isolated discoveries regarding the ellipsoidal figures, like those of Maclaurin and Jacobi, occurred over a long period of time.[1] Beyond these sporadic events we may identify two principal periods of the development of the theory of the ellipsoidal figures and their stability. The first of these occurred in the middle of the last century and was initiated by Dirichlet, and the second began toward the end of that century and was initiated by Poincaré and Lyapunov.

7.2.1 Dirichlet's problem

In his lectures on partial differential equations for the term 1856–7, Dirichlet included a description of certain solutions he had found of the equations of inviscid fluid dynamics. He had begun to write these up into a coherent whole, but his untimely death prevented the completion of this project. The completion was left to Dedekind, who not only put together the completed sections that Dirichlet had left, but also organized scattered notes into further sections of the paper, and followed the completion of Dirichlet's paper with a paper of his own, containing what is now called Dedekind's theorem. These papers (Dirichlet 1860; Dedekind 1860) are published consecutively in the same issue of the *Journal für die Reine und Angewandte Mathematik* of 1860. Riemann's paper (Riemann 1861), published the following year in the *Abhandlungen der Königlichen Gesellschaft der Wissenschaften zu Göttingen*, reorganizes Dirichlet's solution into a somewhat different form (closer to that found in *EFE*), and discusses the steady state solutions and their stability to disturbances leaving them ellipsoidal. Subsequent research followed Riemann's presentation, and these ellipsoidal solutions (especially the steady-state solutions) are now usually referred to as the *Riemann ellipsoids*. However, the important mathematical observations underlying the existence of these solutions are due to Dirichlet, as Riemann himself emphasizes. These observations are two.

The first is that a velocity field that is a linear expression in the Cartesian coordinates "linearizes" the equations of fluid dynamics in a sense described below. Dirichlet made a point of using the *Lagrangian* form of the fluid-dynamical equations to introduce this form of the velocity field.[2] In the more familiar *Eulerian* form of the equations of fluid dynamics, the effect of this assumption is that the nonlinear advective term then also becomes a linear expression in the Cartesian coordinates. It is in this sense that Dirichlet's assumption linearizes the equations: if there were no non-

---

[1] The monograph by Lyttleton (1953) contains a more extensive historical development of these figures, including an account of the role they played in the fission theory.

[2] Dirichlet's argument for preferring the Lagrangian form in this context makes very interesting reading.

linear forcing terms present, the equations of fluid dynamics (eqs. 7.2 and 7.3 below) would become linear in the Cartesian coordinates, and one could immediately solve them to obtain a finite system of ordinary differential equations. The idea of linearizing the equations of fluid dynamics through such an assumption has been rediscovered repeatedly (e.g., Craik 1989). The second observation is that the self-gravitational force inside an ellipsoid of uniform density is also given by a linear expression in the Cartesian coordinates. Putting these observations together, and taking due notice of the conditions at the free boundary and the assumption of incompressibility, Dirichlet was led to a system of ordinary differential equations governing the parameters of the system (the semiaxes a_1, a_2, a_3 of the ellipsoid, and six parameters characterizing the velocity field). While the velocity field is linear in the Cartesian coordinates, the differential equations governing the parameters are nonlinear. The Maclaurin and Jacobi families, which are in equilibrium in a rotating reference frame, form a small subclass of solutions of this system.

Dirichlet's problem thus provides a physically meaningful context wherein a daunting system of partial differential equations is reduced to a system of ordinary differential equations of finite order (in the general case, of order twelve). Dirichlet had applied his equations already in 1857, to the following problem (Dirichlet 1860). For the Maclaurin spheroids, it had been observed earlier that there is a maximum angular velocity. That is, if the density is prescribed and one considers a sequence of Maclaurin spheroids of increasing angular momentum (and therefore of increasing eccentricity of meridional cross-section), the angular velocity of the figure at first increases, but subsequently decreases, a maximum occurring at a certain critical value of the angular momentum (cf. Chandrasekhar 1969, p. 79). On the other hand, one can consider a spheroidal fluid mass of the prescribed density whose initial velocity is that of pure rotation with an angular speed exceeding that which is possible for a Maclaurin spheroid. What is the dynamical outcome of these initial data? This question can be addressed in the context of Dirichlet's equations (with the result that the object performs an oscillatory motion).

In the course of editing Dirichlet's notes for publication, Dedekind observed a certain reciprocity in the system of equations, which can be explained in the following way. The fluid velocity consists of two parts: an angular velocity of rigid-body rotation, and a motion of uniform vorticity superimposed on the latter. Each of these motions can be characterized by a three-component vector (time-dependent, in the general case). Interchanging these vectors provides a different solution of the equations for which the geometric figure is the same (i.e., the semiaxes of the ellipsoid are identical in the two motions). In *EFE* these two motions are said to be *adjoint* to each other, since they are obtained by taking the transpose, or adjoint, of a certain matrix. An example of such a pair of adjoint configurations is the Jacobi-Dedekind pair: the Jacobi ellipsoid is at rest in a frame of reference rotating about the z-axis with angular speed ω, and the Dedekind ellipsoid is at rest in the inertial frame but with a fluid velocity of constant vorticity $\zeta = -\omega \left(a_1^2 + a_2^2\right)/a_1 a_2$.

Riemann rederived Dirichlet's equations in a more symmetrical form. His derivation of Dedekind's reciprocity law consists of a single remark. He went further, however, than merely giving a more compact formulation than his predecessors. He also considered quite generally the family of equilibrium solutions of the system of ordinary differential equations (which correspond to steady-state solutions of the Euler equations of fluid dynamics).[3] These he found to be divided into two kinds: those for which the angular velocity and vorticity are aligned along a principal axis of the ellipsoid, and those for which the latter is not true but these vectors lie in a principal plane. He then used the system of ordinary differential equations to study the stability of these steady-state solutions to disturbances of the fluid mass leaving it an ellipsoid. For this he employed a variation of Lagrange's minimum-energy method. The parameter space is two-dimensional: the ratios of semiaxes, say $\alpha = a_2/a_1$ and $\beta = a_3/a_1$, may be chosen as parameters. Then the part of the parameter space occupied by steady-state solutions of the kind considered is a certain region in the $\alpha\beta$-plane. Riemann's method led to the identification of critical points, or critical curves, separating stable from unstable subregions of this region of parameter space.

He further noted that it was feasible to generalize the stability theory by subjecting the Euler equations to initial data representing arbitrary (hence in general nonellipsoidal) disturbances of the steady-state solutions, since doing so would lead to linear partial differential equations, although he did not pursue this himself.

7.2.2 The fission theory

The occurrence of multiple systems of objects in the sky is the most obvious feature of the solar system, but is by no means limited to the solar system. For example, about half the stars in the sky are double stars. How these and other multiple systems form is a continuing issue of current research. A clear statement of an idealized mathematical problem bearing on this issue appears in the classic dynamics text by Thomson and Tait (1879). The underlying idea is that a rotating, self-gravitating fluid mass, initially symmetric about the axis of rotation (like a Maclaurin spheroid),

---

[3]The term *equilibrium* is somewhat ambiguous. Equilibrium solutions of Dirichlet's system of ordinary differential equations correspond to steady-state solutions of the Euler equations of fluid dynamics, which, with few exceptions, are not true equilibria, or even relative equilibria, of that system. We will live with this ambiguity, as in the title of Chandra's book *EFE*.

can undergo an axisymmetric evolution in which it first loses stability to a nonaxisymmetric disturbance, and continues for a while evolving along a nonaxisymmetric family (like the Jacobi family) toward greater departure from axial symmetry; then it undergoes a further loss of stability to a disturbance tending toward splitting into two. These authors made various plausible conjectures regarding this *fission theory* in the context of the known, rigidly rotating figures of Maclaurin and Jacobi.

The problem of fleshing out the mathematical skeleton constructed by Thomson and Tait was taken up independently by Lyapunov (1884) and by Poincaré (1885). Their mathematical treatment of this problem went beyond the particulars of the astronomical problem and laid the groundwork for the area of nonlinear analysis known today as *bifurcation theory*. In the context of the rigidly rotating figures of Maclaurin and Jacobi, the most relevant perturbations of these figures appeared to be those associated with deformations of the free surface described via ellipsoidal harmonics of orders two and three. Ellipsoidal harmonics of order two are, in the limit of linear disturbances, of the kind envisaged by Riemann: the disturbed figure remains an ellipsoid. Ellipsoidal harmonics of order three or higher are not of this kind, and the corresponding analysis carried out by Poincaré and Lyapunov is significantly more complicated for disturbances of this kind.

The outcome of these mathematical analyses did not fully confirm either the speculations of Thomson and Tait or the further speculations of Poincaré, and despite subsequent efforts and clarifications by Jeans (1917), Cartan (1928), and others, the issue of the viability of the fission theory remains unsettled to this day. From the standpoint of the mathematical analysis of the classical ellipsoids, the advances consisted of determining the stability of the Maclaurin and Jacobi figures to certain higher-harmonic disturbances: arbitrarily high in the case of the Maclaurin figures (Bryan 1889; Cartan 1928), through fourth harmonics in the case of the Jacobi figures (Appell 1921).

Thus Riemann's remark, that one could determine the stability of the *more general* class of ellipsoids discovered by Dirichlet to *arbitrary* disturbances remained only partially explored even for the small subclass of figures represented by the Maclaurin and Jacobi figures, and essentially unexplored where the more general Riemann ellipsoids were concerned.

This was the state of the subject at the time when Chandra ran across Bassett's account of it.

7.3   The virial method

The virial theorem has long been used in mechanics to obtain estimates of dynamical motions of systems of particles. It is obtained by taking the scalar product of either side of the force-balance equation with the position vector of the jth particle and summing:

$$\sum_{j=1}^{n} m_j \mathbf{x}_j \cdot \frac{d^2 \mathbf{x}_j}{dt^2} = \sum_{j=1}^{n} \mathbf{x}_j \cdot \mathbf{f}_j (\mathbf{x}), \qquad (7.1)$$

where n is the number of particles. After some elementary manipulations and (possibly) the introduction of plausible assumptions, it provides a relation between the inertial terms on the left and the forcing term on the right. It is known to astronomers in particular for estimating the relative importance of gravitational and inertial effects in groups of stars (as in Ambartsumian's book [1958]).

Its application in fluid dynamics was considered by Rayleigh (1903) and more recently by Parker (1957). To carry out this application in the case of an ideal fluid, one considers the force-balance equation (the *Euler equation*):

$$\frac{Dv_i}{Dt} = -\frac{1}{\rho} \frac{\partial p}{\partial x_i} + f_i (\mathbf{x}, t), \ i = 1, 2, 3. \qquad (7.2)$$

Here $v_i = v_i (\mathbf{x}, t)$ is the i-th component of the velocity, p the pressure, and f the force per unit mass. The operator D/Dt is given by the formula

$$\frac{D}{Dt} = \frac{\partial}{\partial t} + \sum_{i=1}^{3} v_i \frac{\partial}{\partial x_i}.$$

These equations have to be supplemented by others to form a closed system of equations. For the sake of definiteness we suppose the fluid to be incompressible: $\rho = $ constant. This imposes the further condition that

$$\text{div } \mathbf{v} = \frac{\partial v_1}{\partial x_1} + \frac{\partial v_2}{\partial x_2} + \frac{\partial v_3}{\partial x_3} = 0, \qquad (7.3)$$

and the system of equations (7.2) and (7.3) is then closed. It still needs to be supplemented by appropriate initial and boundary conditions. If one multiplies the i-th equation of the system (7.2) by x_i, sums on i from 1 to 3, and integrates over the domain D occupied by fluid, one obtains the analog of the virial theorem for a collection of particles. This clearly results in a relation among integrals involving inertial terms (from the left-hand side of the equation) and terms involving forces, due to fluid pressure and whatever further forcing terms are present (from the right-hand side of the equation). This relation is called the virial equation (or *scalar* virial equation).

It is not *a priori* evident that the relation obtained in this way will be a useful one. However, if you grant that it may be useful, there are immediate generalizations of it that may then also be useful. Instead of multiplying the i-th equation by x_i and summing, one can multiply the

j-th equation by x_i and have two free indices i, j, providing nine equations in all:

$$\int_D x_i \frac{Dv_j}{Dt} \rho dV =$$
$$- \int_D x_i \frac{\partial p}{\partial x_j} dV + \int_D \rho x_i f_j (x, t) \, dV, \quad i, j = 1, 2, 3, \qquad (7.4)$$

or, after manipulating the left-hand side,

$$\frac{d}{dt} \int_D x_i v_j dV - \int_D v_i v_j \rho dV =$$
$$- \int_D x_i \frac{\partial p}{\partial x_j} dV + \int_D \rho x_i f_j (x, t) \, dV, \quad i, j = 1, 2, 3. \qquad (7.5)$$

The condition (7.3) must also be taken into account (even in the scalar case). For an ellipsoid of semi-axes a_1, a_2, a_3 it implies the further relation

$$\int_D \left(\frac{x_1 v_1}{a_1^2} + \frac{x_2 v_2}{a_2^2} + \frac{x_3 v_3}{a_3^2} \right) dV = 0. \qquad (7.6)$$

These equations must contain more information than the earlier scalar virial equation since the latter is derivable from them. They form the so-called *tensor* virial equations. One need not stop there: the k-th equation of the system (7.2) can be multiplied by $x_i x_j$ and integrated over D, providing twenty-seven equations in all. And so on.

A paradigm for investigating differential equations describing the evolution of an interesting physical system is: (a) find the steady-state solutions; (b) investigate their stability. Ledoux's application, and the initial investigations employing the virial tensor in the context of the ellipsoids, were in the context of stability of known steady-state figures. For this application of the tensor virial equations one needs these equations, not in the form given by equation (7.5), but in a form derived from the latter by perturbation about a known solution. In other words, one considers equation (7.5) for the known solution, then the same equation but for the unknown (or *perturbed*) solution, and subtracts. The difference may be written conveniently in term of the *Lagrangian displacement* $\xi(x, t)$. The latter is the vector from the position $\mathbf{x}$ of a fluid particle in the unperturbed flow to the position of the same fluid particle for the perturbed flow. A knowledge of the Lagrangian displacement as a function of position and time provides a complete description of all the flow variables in the present, conservative context. If we define the variables

$$V_{i;j} = \int_D \rho \xi_j x_i dV \text{ and } V_{ij} = V_{i;j} + V_{j;i},$$

we can express the tensor virial equations as follows (see *EFE* for details):

$$\frac{d^2}{dt^2}V_{i;j} + \frac{d}{dt}\int_D \rho\left(\xi_j v_i - \xi_i v_j\right)dV =$$

$$\delta\mathcal{T}_{ij} + \delta\mathcal{W}_{ij} + \omega^2 V_{ij} - \omega_i\omega_j V_{kj} + \delta_{ij}\delta\Pi$$

$$+ 2\epsilon_{ilm}\omega_m\delta\int_D \rho v_i x_j dV, \qquad (7.7)$$

where δ indicates the difference between the perturbed and unperturbed version of the expression following it. Here double indices not separated by a semicolon are symmetric in their indices. The terms in $\delta\mathcal{T}$, $\delta\mathcal{W}$, and $\delta\Pi$ refer to quantities involving kinetic energy, potential energy, and pressure respectively. The first two of these can be expressed in terms of the variables V_{ij}. The incompressibility condition can be shown to imply that

$$\frac{V_{11}}{a_1^2} + \frac{V_{22}}{a_2^2} + \frac{V_{33}}{a_3^2} = 0. \qquad (7.8)$$

For an incompressible ellipsoid, these equations, nine in number after the term involving the pressure is eliminated, represent a homogenous system in the nine unknowns. Thus no ansatz is required. At least some of these nine quantities are nonzero if the surface deformation of the ellipsoid is given by a second-order ellipsoidal harmonic. Hence, among the solutions for the oscillation frequencies are those belonging to the second-harmonic perturbations of the ellipsoid. To achieve perturbations of the ellipsoid given by third harmonics, one requires the higher-order virial equations obtained by taking moments of the ith equation with $x_j x_k$. And so on.

The advantages of these equations in comparison with methods employed earlier, involving expansions in ellipsoidal harmonics, may be explained as follows. The use of the latter is arduous in part because the ellipsoidal coordinates suffer from a feature not shared by the more widely used coordinate systems (like Cartesian and spherical coordinates): they are not a single coordinate system but a parameterized family of coordinate systems depending on the semi-axes of the particular ellipsoid. Correspondingly, the ellipsoidal harmonics form not a single complete system of functions but a parameterized family of complete systems: they have to be calculated anew for each ellipsoid under investigation. In using expansions in ellipsoidal harmonics, one uses (explicitly or implicitly) the orthogonality properties of these functions when integrated over the fundamental domain to project a function onto some finite-dimensional subspace. Likewise with spherical harmonics. For expansions in Cartesian coordinates, the virial method turns out to play the role of the projection procedure. It has the advantage that the "harmonics," which are simply the monomials in powers of the Cartesian coordinates, are known once and for all and do not depend on the object under investigation. It has the disadvantage as well that it is necessary to work out a different set of virial equations corresponding to each order of ellipsoidal harmonics. Equation (7.7) above corresponds to second-order harmonics. The system that would be obtained by multiplying the kth equation by the arbitrary monomial $x_i x_j$ corresponds to third-order harmonics. These systems rapidly become unwieldy and are limited for most practical purposes to low orders.

It is also true in the context of the incompressible ellipsoids that the equations of the unperturbed flow are given by the tensor virial equations (7.5), in the following sense. If one substitutes into these equations the structure of Dirichlet's solution with unknown parameters, the tensor virial equations then determine the relations among the parameters.

There is no link in principle between the virial equations on the one hand and the classical ellipsoids on the other. The virial equations can be formulated for arbitrary kinds of fluid configurations, with corresponding changes in the forms they take. They are, however, particularly well adapted to the study of the ellipsoids.

7.4   Chandra's contributions

The work of Chandra and his collaborators broke "old ground" as well as "new ground." The old ground consisted of applications of the virial method to problems concerning the Maclaurin and Jacobi families that had been considered earlier, especially within the context of the fission theory.[4] The new ground consisted of novel applications, not only to the Riemann ellipsoids but also to other fluid-dynamical problems. Some of these contributions are now summarized.

7.4.1   Old ground

The early applications were to the Maclaurin and Jacobi figures (recall that Chandra did not know of the more general figures of Riemann until 1964). The oscillation frequencies of the Maclaurin figures were calculated for perturbations associated with second- and third-order ellipsoidal harmonics. The location of bifurcation points under surface deformations described by third-harmonics along the Maclaurin and Jacobi families was also carried out using the virial equations. Since the locations of these points were either already known or deducible on the basis of already established technique, one may ask why one should do them again. There are complemetary reasons for this.

Recalculating the classical bifurcation points from the new standpoint reconfirms the older results from a computationally distinct viewpoint, val-

---

[4] Chandra chose not to address the fission theory directly. To do so would have involved a heavy investment in nonlinear bifurcation analyses, whereas he was more interested in exploring the capabilities of the method in linear theory.

idates the new procedure, and provides computational experience with the new technique that will be needed in breaking new ground. There had been ample confusion regarding the interpretation of the classically calculated bifurcation points (cf. Lyttleton 1953, chap. 1, regarding this), and therefore scope for reconfirmation. Computational experience with the new technique was important in order to take advantage of the virial method. This preliminary series of investigations showed very convincingly that one could indeed find all the critical points that had been found classically with an essentially elementary technique, i.e., without ever constructing, or even explicitly introducing, the ellipsoidal harmonics.[5]

7.4.2 New ground

The first novel applications of the virial method were to compressible rather than incompressible masses. For this application a variant of equation (7.7) is needed. These equations required a special development to handle the gravitational terms, leading to the *superpotentials*, scalar quantities generalizing the gravitational potential. These developmental matters were attended to in a series of papers (cf. Chandrasekhar and Lebovitz 1962a,b). The equations could then be applied in specific contexts (including rotating polytropes, for example: cf. Chandrasekhar and Lebovitz 1962c). An application of the results to a concrete astronomical problem was the interpretation of the beat period of the β *Canis Majoris* stars (Chandrasekhar & Lebovitz 1962d, e, f). The compressible theory for a spherical star indicated that, for a critical value $\gamma = 1.6$ of the ratio of specific heats, the fundamental mode of radial pulsation and the P_2 mode of nonradial pulsation were degenerate (i.e., have the same frequency). A small rotation would lift the degeneracy, and neither of the two resulting normal modes was radially symmetric. The result is that, under the influence of an essentially spherical forcing, both modes would be excited with comparable amplitudes, resulting in a steady beating with a frequency given by the difference of the two characteristic frequencies.[5]

The rediscovery of the Riemann ellipsoids opened extensive new ground for the application of the method. However, the first step was again intended to be "old ground": Riemann, in his paper of 1861, had discussed the stability of the equilibrium solutions that he had found to perturbations leaving them ellipsoidal. This made it possible to consider stability in the context of the ordinary differential equations describing the ellipsoidal motion. He described his conclusions by giving the neutral curves in the space of the parameters $\alpha = a_2/a_1$ and $\beta = a_3/a_1$; these represent figures on the borderline of instability, separating stable subregions of the

---

[5] This interpretation of the beat phenomenon was received coolly by the larger community of astronomers.

parameter space from unstable subregions. Chandra set out to confirm this with the aid of the tensor virial equation (7.7). What he found for stability boundaries agreed in some domains of parameter space, but showed discrepancies in others. The pattern of discrepancy was such that wherever Riemann concluded stability Chandra agreed, but there were small regions where Riemann concluded instability but Chandra concluded stability. The natural inclination to defer to the great German mathematician conflicted with a careful reexamination of both Chandra's own methods and Riemann's. Riemann did not calculate sets of oscillation frequencies, but rather used a version of *Lagrange's theorem*: he found a function constant on orbits, the vanishing of whose gradient gives the equilibrium conditions, and associated stability with minima of this function. However, the converse association of instability with critical points that fail to be minima was not justified. In Lagrange's theorem the latter association is justified because the conserved quantity is the sum of a positive-definite kinetic energy and a potential energy. Riemann's conserved quantity takes this form only for a subfamily of his ellipsoids (the S-type ellipsoids, defined below), and here his stability conclusion is in exact agreement with that of Chandra's virial analysis. The pattern of discrepancy is consistent with a misapplication of Lagrange's theorem (Lebovitz 1966). Hence what was to have been old ground opened new ground instead, correcting aspects of Riemann's analysis.

The problem of the oscillations and the stability of the Jacobi family is of particular significance, since the point along that family where instability sets in played a major role in the fission theory. Cartan (1928) had shown that the Jacobi family becomes dynamically unstable at the point of bifurcation along this family originally isolated by Poincaré (1885), from which the *pear-shaped* family bifurcates. He had not, however, explicitly calculated the oscillation frequencies of the Jacobi family under perturbation by the associated third-harmonics. These frequencies, found in detail via the virial technique (Shore 1963), confirm Cartan's theorem in a graphic manner.

The *S-type ellipsoids*, already referred to above, are a subfamily of the Riemann ellipsoids for which the angular velocity and vorticity are directed along the same line (the z-axis, say). Chandra also considered their stability to third-harmonics disturbances, but only for neutral disturbances (i.e., oscillation frequencies were *not* calculated). This enabled a generalization to this family of ellipsoids of the analysis of Poincaré for the Jacobi family, in keeping with the intention of bringing the study of the Riemann ellipsoids to the level of completion that had previously been achieved for the Jacobi family.

Another important mathematical element of the fission theory of binary stars was the assertion, by Thomson and Tait (1879), that the Maclaurin spheroids would become (secularly) unstable to an ellipsoidal disturbance at the point where the Jacobi family bifurcates from it if dissipation is present, and not otherwise, i.e., that dynamical instability does not set in at this point. The latter point had been explicitly demonstrated, but the former had not. While the reasons given by Thomson and Tait were sound and generally accepted, an explicit confirmation was presented only in the 1960s (by Roberts and Stewartson [1963] and by Rosenkilde [1967]). Rosenkilde's approach was to use the virial theorem for a viscous liquid with an ansatz for the Lagrangian displacement drawn from the inviscid theory. The two approaches give the same result, fully confirming the assertion of Thomson and Tait. But Rosenkilde's approach is remarkable for its simplicity.

Chandra also considered problems in which tidal forces join with rotation to determine the shape of the free surface, with the approximation of the tidal force such that the figures are ellipsoids. One of these, the Roche problem, envisages a liquid figure tidally distorted by a point mass. Here again the issue of secular stability arises, and again this issue was settled in a remarkably simple fashion by Rosenkilde's method.

As I have mentioned, Chandra resented some of the time spent on the ellipsoidal figures because of his eagerness to continue his work in relativity. His experience with the ellipsoids served him well, however, in his subsequent research on gravitational radiation in the post-Newtonian approximation. Here he found (Chandrasekhar 1970) a useful paradigm in the adjoint Jacobi and Dedekind figures, the first radiating because its figure is rotating in an inertial frame, the second not radiating because its figure is at rest in an inertial frame. A now-standard and widely quoted reference on dissipation through the effects of gravitational radiation and of viscosity is the work of Detwiler and Lindblom (1977), which takes as its point of departure the theory of the S-type Riemann ellipsoids.

7.5   A retrospective view

During the period when he worked on the classical ellipsoids, Chandra endured criticism from a number of astronomers, many of whom felt that the ellipsoids were not relevant to the mainstream problems of astronomy.[6] Chandra himself expressed impatience with the subject from time to time. One can therefore fairly inquire what the outcome of this intense research activity has been, what lasting influence it has had, in astronomy in particular and in science more generally. Indeed he addressed these questions himself in an epilogue to *EFE*. There he limits himself to two remarks: (1) this physically motivated and mathematically beautiful subject had been badly neglected, and it seemed a pity to leave it in such a neglected

---

[6] This attitude of his colleagues was not restricted to this period of his work: Chandra has said that he was never part of the astronomical establishment.

condition, and (2) he wanted to give a substantial exposition of the virial method, which has applications beyond the classical ellipsoids. I would add a third goal, however, which he expressed personally, regarding the difficult and time-consuming procedure of writing *EFE*: he felt that, if he did not make this effort, the classical ellipsoids and the preceding efforts over a period of almost nine years would largely be forgotten. Now, some three decades later, we can perhaps address the success of these goals.

Regarding the popularization of the virial method, success has been modest. There indeed has been a stream of applications over a long period, and the stream, while never a roaring current, does not seem to be dying out. The form of the virial theorem found in current textbooks (cf. Binney and Tremaine 1987; Shore 1992) is the tensor form. However, its extensive application supplanting expansions in harmonics has not caught on. There are reasons for this. One is that it has the disadvantage, mentioned above, that a different set of virial equations has to be defined at each order, and these become increasingly cumbersome at higher order. Furthermore, many users of *EFE* refer to Chandra's technique as "sophisticated." He intended exactly the opposite! However, the technique, while elementary in the sense of not requiring a knowledge of ellipsoidal harmonics, nevertheless requires its own specialized development, which is not part of a standard scientific education.

There is, however, one particular success of the method that has never been followed up and fully explained. This is what I have called "Rosenkilde's method," for flows of low Reynolds number (i.e., small viscosity). The standard approach to estimating the effect of a small viscosity is boundary-layer theory, and it can lead to very heavy calculations (cf. Roberts and Stewartson 1963), while Rosenkilde's method is extremely simple and elegant by comparison. Its success must be related to the circumstance that the underlying equations incorporate the exact, viscous boundary conditions while the ansatz introduces the "outer solution" of boundary-layer theory. However, neither the details of this correspondence nor the limits of the method's validity have been adequately explored.

Regarding the neglected state of the subject of the Riemann ellipsoids, Chandra's efforts clearly accomplished significant restoration. The stability to second harmonics was considered *ab initio* and Riemann's conclusions corrected. Bifurcation points under third-harmonic disturbances were worked out for the S-type ellipsoids. There are isolated cases where the same is done under fourth-harmonic disturbances. This brought the subject to a level of completion similar to that which had previously existed for the Maclaurin and Jacobi figures. However, a complete study of the dynamics of the Riemann ellipsoids was not, and has not yet been, achieved (although further progress has in fact been made recently; see below). Riemann's remark, that the study of the stability of these figures leads "only to linear differential equations," now sounds rather innocent in view of the effort needed to make progress in the subject.

The goal of writing the book, to prevent the subject of the Riemann ellipsoids and the advances Chandra and his collaborators had made in it from disappearing from the scientific scene, has succeeded admirably. That *EFE* has become the principal reference on the classical ellipsoids is of course true, but this statement doesn't go very far, since *EFE* is the *only* extensive reference. It is, however, further true that the book has brought these figures to the attention of astronomers, physicists, and mathematicians (to name those areas in which I have personal knowledge of research activity), allowing applications to be made that might not have occurred to their authors if there had been no such book. Many of the applications in astronomy (where real stars do not conform to the rigid hypotheses of the theory of the ellipsoids) and in physics (where the liquid-drop model of the nucleus involves figures only approximately ellipsoidal and involving Coulomb forces and surface tension rather than gravity) have an approximate character. For mathematicians, the Riemann-Dirichlet equations represent a rich Hamiltonian system harboring a variety of behaviors. For all of these, *EFE* is a well-known, well-written, and easily accessible guide to the subject.

Chandra's pattern of writing a book and moving to a new subject has sometimes intimidated those who wished to work in the field he just left: there is concern that everything worth doing has been done. His reason for establishing this pattern was quite different: he wanted to state what he had learned of the subject in a coherent form. This should be a help, rather than a hindrance, to those who wish to study the subject further, and indeed it has been. I'll conclude with three recent examples of research activity extending our understanding of this area of science which Chandra resurrected.

One area that was clearly not exhausted by Chandra is that of the stability of the Riemann ellipsoids. One recent development has been a reconsideration of the stability of the S-type ellipsoids (not via the virial method, but with the aid of the ellipsoidal harmonics and some help with symbol-manipulation computer programs). Oscillation frequencies have been calculated for disturbances up to fifth harmonics, and have been complemented by a WKB analysis for arbitrarily small wavelengths (Lebovitz and Lifschitz 1996a,b). These reveal fluid-dynamical instabilities associated with the strain component of the velocity field (rather than with the energetics associated with the gravitational and rotational fields).[7] These previously undetected instabilities affect most of the parameter space, and have rather large growth rates for the Dedekind family and nearby figures, which are characterized by large strain. This is not the place to speculate

---

[7] Indeed, the stability conclusions for these figures, even for third-harmonics disturbances as given in *EFE*, require revision.

on the implications these new results have for the applications of these classical figures.

Another recent development is the discovery (Marshalek 1996) of a limiting form of Riemann ellipsoids *not* of type S. This is an irrotational family of figures whose angular velocity does not lie along an axis but in a principal plane, overlooked by Riemann and not pointed out in *EFE*. It has similarities with the *tilted rotor* model of recently discovered atomic nuclei.

Finally, in Darwin's tidal problem, it has been pointed out (Lai *et al.* 1994) that if, instead of using an approximation to the tidal potential making the figure exactly an ellipsoid, one uses a variational principle in which the linear velocity field appears in the form of a trial function, the restriction of Darwin's tidal problem to congruent masses can be relaxed, and an improved formula for the angular velocity be obtained, consistent with arbitrary masses for the two components.

References

Ambartsumian, V. A. 1958, *Theoretical Astrophysics* (London: Pergamon Press), 525.

Appell, P. 1921, *Traité de Mécanique Rationelle*, vol. 4 (Paris: Gauthier-Villars et Cie).

Bassett, A. 1888, *A Treatise on Hydrodynamics*, vol. 2 (Cambridge: Deighton, Bell and Company; reprinted edition 1961, New York: Dover Publications).

Binney, J., Tremaine, S. 1987, *Galactic Dynamics* (Princeton: Princeton University Press).

Bryan, G. H. 1889, *Phil. Trans.*, **A180**, 187.

Cartan, H. 1928, *Proc. Int. Math. Congress, Toronto, 1924*, **2** (Toronto: University of Toronto Press).

Chandrasekhar, S., Lebovitz, N. R. 1962a, *Astrophys. J.*, **135**, 238.

Chandrasekhar, S., Lebovitz, N. R. 1962b, *Astrophys. J.*, **136**, 1032.

Chandrasekhar, S., Lebovitz, N. R. 1962c, *Astrophys. J.*, **136**, 1082.

Chandrasekhar, S., Lebovitz, N. R. 1962d, *Astrophys. J.*, **135**, 305.

Chandrasekhar, S., Lebovitz, N. R. 1962e, *Astrophys. J.*, **136**, 1069.

Chandrasekhar, S., Lebovitz, N. R. 1962f, *Astrophys. J.*, **136**, 1105.

Chandrasekhar, S., Lebovitz, N. R. 1964, *Astrophysica Norvegica*, **IX**, 323.

Chandrasekhar, S. 1969, *Ellipsoidal Figures of Equilibrium* (New Haven: Yale University Press; reprinted edition 1987, New York: Dover Publications).

Chandrasekhar, S. 1970, *Phys. Rev. Lett.*, **24**, 611.

Craik, A. 1989, *J. Fluid Mech.*, **198**, 275.

Dedekind, R. 1860, *J. Reine und Angew. Math.*, **58**, 217.

Detwiler, S., Lindblom, L. 1977, *Astrophysical J.*, **213**, 193.

Jeans, Sir James 1917, *Mem. R. Astron. Soc. London*, **62**, 1.

Lai, D., Rasio, F., Shapiro, S. 1994, *Astrophys. J.*, **423**, 344.

Lebovitz, N. R. 1961, *Astrophys. J.*, **134**, 500.

Lebovitz, N. R. 1966, *Astrophys. J.*, **145**, 878.

Lebovitz, N. R., Lifschitz, A. 1996a, *Astrophys. J.*, **458**, 699.

Lebovitz, N. R., Lifschitz, A. 1996b, *Phil. Trans.*, **A354**, 927.

Ledoux, P. 1945, *Astrophys. J.*, **102**, 143.

Lejeune Dirichlet, G. 1860, *J. Reine und Angew. Math.*, **58**, 181.

Lyapunov, A. 1884, thesis (Kharkov); French translation (1904) in *Ann. Fac. Sci. Univ. Toulouse, ser. 1*, **6**.

Lyttleton, R. A. 1953, *The Stability of Rotating Liquid Masses* (Cambridge: Cambridge University Press).

Marshalek, E. R. 1996, private communication.

Parker, E. N. 1957, *Astrophysical J. Suppl.*, **3**, 51.

Poincaré, H. 1885, *Acta. Mathematica*, **7**, 259.

Rayleigh, Lord 1903, *Scientific Papers*, vol. 4 (Cambridge: Cambridge University Press), 491.

Riemann, B. 1861, *Abh. d. Königl. Gesell. der Wis. zu Göttingen*, **9**, 3.

Roberts, P. H., Stewartson, K. 1963, *Astrophys. J.*, **137**, 777.

Rosenkilde, C. E. 1967, *Astrophys. J.*, **148**, 825.

Shore, S. N. 1992, *An Introduction to Astrophysical Hydrodynamics* (San Diego: Academic Press), 12.

Shore, S. N. 1963, *Astrophys. J.*, **137**, 1162.

Thomson, W. and Tait, P. G. 1879, *Treatise on Natural Philosophy* (Cambridge: Cambridge University Press).

8

Making the Transition from Newton to Einstein: Chandrasekhar's Work on the Post-Newtonian Approximation and Radiation Reaction

Bernard F. Schutz

8.1 Chandrasekhar the relativist

Chandra first became interested in general relativity in the early 1930s. He had already discovered that white dwarfs had a maximum mass, and Eddington had pointed out that this would imply that stars of a larger mass could collapse to black holes.[1] Eddington thought this idea so abhorrent that he felt forced to dismiss Chandra's work on white dwarfs; this story is described elsewhere in this volume. But Chandra drew the opposite conclusion. He realized that he would need to understand general relativity in order to follow the implications of his discovery to their natural conclusions (Chandrasekhar n.d.).

Nevertheless, he did not study relativity immediately. Discouraged by the evident hostility towards general relativity shown by many prominent physicists in the 1930s (in his private memoirs (Chandrasekhar n.d.) he mentions Bohr in particular), and believing that general relativity had already proved to be a "graveyard of many theoretical astronomers," he steered a different course. In the 1950s, when he again thought of the subject, he remarked that astronomers doing relativity "were prone to play for high stakes," while his own "approach to science was more conservative"; this perception was later to be borne out by his work on the post-Newtonian

---

[1] The term "black hole" was, of course, not used by Eddington; it was coined by John Wheeler in the 1960s.

approximation, and in particular in deriving the reaction effects of gravitational radiation, where his conservatism helped him avoid mistakes that had been made by others. He finally took up his interest in relativity in the 1960s, when he had such a strong scientific reputation that, as a friend said to him, "What can you lose?"

But once he started in relativity, he never left it. After the 1960s, Chandra worked almost exclusively on problems in general relativity. Even in the 1960s, his impatience to finish other research projects and get on with relativity is evident in his memoirs. He repeatedly refers to "distractions" that prevent him from working full-time on relativistic problems; among these distractions are, surprisingly, some of his best-known work, such as the last stages of his work on the figures and stability of rotating homogeneous bodies in Newtonian gravity, and completing his book *Ellipsoidal Figures of Equilibrium* (Chandrasekhar 1969). His heart was in general relativity, which he had characteristically been learning by teaching courses on it while he finished up his other interests, and he was impatient to get to grips with new problems in it.

The first problem Chandra chose was to study the pulsation of a spherical star in general relativity. It proved surprisingly tractable, and it immediately had an impact on astronomy in showing that quasars were unlikely to be single supermassive stars. His work on this is described in chapter 9 (by John Friedman) in this volume; and there is an interesting comment on it in Thorne's foreword to one of the volumes of Chandra's collected papers (Thorne 1990).

Having gained confidence from this first project, Chandra wanted a more far-reaching and fundamental goal to shoot for. He chose a problem that had defied solution for more than 40 years: how does the emission of gravitational radiation affect the emitting system, in particular when the system is self-gravitating? Many relativists had studied this problem, called *radiation reaction*. Some had provided partial answers, others had got completely wrong results. The situation was so confused that it had led some to doubt the reality of gravitational radiation, or at least the possibility of associating some kind of conserved energy with gravitational waves. This confusion was very unsatisfactory from both the physical and astrophysical points of view, and Chandra must have seen that there was a huge unexplored territory in astronomy waiting for him if he could clear up the confusion.

It was characteristic of Chandra's approach to physics that he saw this problem as one that he could solve by a careful step-by-step approach. He would start with the Newtonian limit of general relativity and introduce successively higher-order corrections until he came to the place where radiation reaction could be found. He would develop the post-Newtonian expansion to general relativity in complete detail.

No one before Chandra had attempted the post-Newtonian approximation for continuous bodies in such an exhaustive way. It was well known in relativity that the effects of radiation reaction could be expected to manifest themselves at a high order in such an expansion, but essentially every previous approach to finding them used one device or another to simplify the calculation. It was in handling these simplifications that most previous work had gone wrong. Chandra decided not to try tricks, but rather to trust in his ability to carry the calculation to a high order without getting lost in its complexity.

Chandra introduced no major new tools or conceptual breakthroughs in his calculation. Nor were his calculations particularly polished; some of the more awkward features of the work were later to lead him into controversy. Chandra studied previous work carefully, until he knew what methods to emulate and what errors to correct. But essentially his method would have been available to anyone. No one before, however, had attempted to push it all the way to completion. In this he was true to his perception, quoted above, that he approached problems more conservatively than others. Previous work on this problem had aimed high, going for the radiation terms directly in one way or another, and failing. His approach was to keep his sights firmly on algorithmic calculations that he was confident would eventually lead him to the answer.

8.2   Post-Newtonian relativity

8.2.1   Background

I shall not review the history of approximate treatments of the equations of motion here. Excellent reviews can be found in Damour (1987) and the very recent study by Kennefick (1996). But it is useful to discuss briefly some of the simplifications that had gone wrong in earlier attempts at the radiation-reaction problem in relativity.

Radiation reaction is well understood in electromagnetism, where in a low-velocity expansion of the equations of motion of a charged body the reaction effects arise at order $(v/c)^3$. Put another way, the effective reaction force on a radiating charged body is proportional to the third time derivative of its position, or equivalently to the first derivative of its acceleration.[2] All attempts at radiation reaction in general relativity started with lessons learned from the electromagnetic problem. This included using simplifications that worked successfully in electromagnetism.

The most common simplification was to treat the bodies as point particles. This works well in electromagnetism, despite the fact that the electromagnetic self-field of a point charge is infinite on the body, and this self-

---

[2] Time derivatives in a slow-motion approximation raise the order of quantities by one power of v, so three time derivatives of a term like the position of a body produce a term of order $(v/c)^3$.

field contains an infinite self-energy. By either absorbing the self-energy into the mass of the body, or using the radiative Green function introduced by Dirac (1938) (which is a trick that I will return to in the next paragraph), it is possible to show that radiation reaction acts on a point charge *only* at third order (Jackson 1962); there are no higher-order corrections. When the point-particle method was tried in general relativity, however, it ran into troubles with nonlinearity. The infinite self-energy creates higher-order corrections to the Newtonian gravitational field that themselves get arbitrarily strong near the point mass, and handling these is delicate. In general relativity, it turns out to be possible to develop the post-Newtonian approximation by assuming the body is compact, and treating the gravitational field outside it as if the body were a point mass.

In fact, Fock (1959), Peres (1960), and Peters (1964) had obtained the correct reaction formula in essentially the correct ways, but because the point-particle approach had also yielded incorrect answers in other hands, their papers were not universally accepted.[3] Transparently successful ways of doing this were eventually developed by, among others, Damour (1983), Walker and Will (1980), Walker (1984), and Anderson (1975), but these came after Chandra's work.[4]

The other simplification was to try to skip intervening orders in the expansion and calculate only the radiation-reaction terms. This is a trick that is closely related to Dirac's method for the point charge. In general relativity, as in electromagnetism, the fundamental equations are invariant under timereversal, so that any dissipative effect, like the loss of energy to radiation, must be put in explicitly in some way. The conventional point of view[5] is that the field that one must use is the particular solution of the field equations that has no radiation coming into the system from the distant past; this is the solution that, in a linear theory like electromagnetism, follows from the retarded Green function. In nonlinear general relativity, one effectively puts in time-asymmetry by adopting a boundary (or asymptotic) condition of no incoming radiation on the field.

This condition is easy to implement in a linear field theory, such as electromagnetism or in linearized gravitation. In linearized theory, radiation reaction turns up in a small-velocity expansion of the metric tensor (its 00 component) at order $(v/c)^5$.[6] Its implementation in a nonlinear

---

[3] In particular, Chandra was not aware of Peters's work before he completed his own, and he may not have been aware of Peres's work either.

[4] Damour began his work completely unaware of Chandra's calculations; Walker and Will, and Anderson, had studied the full history and tried to correct previous errors.

[5] There are other points of view, for example that the time-asymmetry arises statistically from random initial data (Schutz 1980). This subject is part of the physics of the arrow of time (Landsberg 1982).

[6] It is no mystery that this is at a higher order than in electromagnetism. The electromagnetic radiation reaction force at order v^3 is proportional to $d^3(\int \rho x^i d^3x)/dt^3$, where ρ is the charge density. Because charge is conserved, this is equivalent to

field theory is not so straightforward, and in general relativity this has been a subject of considerable debate (Ehlers 1980). One way to isolate the radiation-reaction terms while ignoring all the others (thereby skipping lower orders) is to write the solution of the field equations using the retarded Green function and then expand it and the equations of motion in powers of v/c, keeping only the *odd* powers. Done correctly, this trick does work, but at the time Chandra began it it had not been done correctly.

A related but more sophisticated version of this trick is to use the method of matched asymptotic expansions to find the terms in the equation of motion that couple directly to the outgoing radiation far away. This method was applied to the problem by the late William R. Burke and his Ph.D. supervisor, Kip Thorne (Burke 1969; Thorne 1969). Interestingly, they began their work after Chandra embarked on the full post-Newtonian calculation, and they reached their goal of radiation reaction (skipping lower orders) a year before he did, largely because of the "distractions" referred to earlier, which frustrated Chandra's progress on the calculation. Nevertheless, Burke himself was not entirely convinced of the correctness of his result until Chandra derived an equivalent result. Many other relativists have viewed the Burke-Thorne method with some uneasiness (Ehlers *et al.* 1976), although in fluid dynamics the method of matched asymptotic expansions is an accepted tool that leads to derivations of important and experimentally testable results.

8.2.2 The post-Newtonian expansion

Against this background, Chandra felt that he would only have confidence in a full, order-by-order calculation. This was inevitably complicated because of the nonlinearity of Einstein's equations. When the system is self-gravitating and basically well-described by Newtonian gravity, orders in nonlinearity are related to orders in velocity. The virial theorem implies that the gravitational potential Φ should be of the same order as v^2, so that reaction terms from linear theory will not be the only $(v/c)^5$ terms. Terms of order $(v/c)^3(\Phi/c^2)$ and $(v/c)(\Phi/c^2)^2$ will be just as large, and must be considered. In addition, it could in principle happen that the nonlinearities would create reaction effects at lower order, say of order $(v/c)\Phi$. This would have to be looked for.

$d^2(\int \rho v^i d^3x)/dt^2$. The same result is true in linear gravity, provided we take ρ to be the mass density. But the integral $\int \rho v^i d^3x$ in this case is the total momentum of the system, which is conserved. (In general relativity, since all energies and momenta radiate, one must consider radiation from the entire mechanical system, not just one piece of it.) Its time-derivatives therefore vanish, and there is no gravitational radiation reaction at this order. At order v^5, the reaction forces in both electromagnetism and linear gravity depend on $d^5(\int \rho x^i x^j d^3x)/dt^5$, with ρ suitably interpreted. This argument was well understood in relativity by the time Chandra began his work.

At the lowest order (zero-order in our way of counting), the expansion starts with the Newtonian equations. What can we say about the post-Newtonian expansion? The first observation is the one we made earlier, namely that Einstein's equations are, like the rest of classical physics, invariant under time reversal. So we expect the equations of motion to expand only in even powers of v, unless we introduce explicitly time-asymmetric terms in, say, the initial conditions or the boundary conditions. An initial non zero velocity will introduce terms linear in v into the metric components, but Chandra was only interested in the equations of motion, and these momentum-dependent metric terms affect the motion only by coupling to the velocity of the body, so that they appear as order v^2 corrections to the equations of motion. So the first nontrivial correction to the equation of motion would be of order $(v/c)^2$, or of order Φ. This is called the *first post-Newtonian* order, denoted 1pN. Post-Newtonian nomenclature numbers the orders according to their order beyond the Newtonian one in a nonlinearity expansion in Φ.

It turns out that there are no nontrivial terms in the equation of motion at order v^3, so the next corrections are at the second post-Newtonian (2pN) order. However, there will be terms of order v^5, where we expect radiation reaction, so this was Chandra's goal: $2\frac{1}{2}$pN order.

8.2.3   Chandra's work

First post-Newtonian order

Chandra began working on the post-Newtonian approximation in spherical symmetry, where the equations are simplest. He found in this way the post-Newtonian dynamical instability of white dwarfs and radiation-dominated stars (Chandrasekhar 1964), but he soon turned to the full equations, without assumptions of symmetry. Chandra began with the 1pN equations of motion for a perfect fluid (Chandrasekhar 1965). He adopted his own choice of gauge, not equivalent to the de Donder gauge that had been used by most workers before him. He chose a gauge in which he could make the equations of motion look as similar to those of the Newtonian case as possible. He showed that the metric and the motion of the body could be adequately described at this order by a few potentials that satisfied generalizations of the Poisson equations. Chandra was particularly concerned to ensure that his post-Newtonian theory was physically reasonable and complete at this level of approximation, so he derived (by inspection) expressions for the energy and angular momentum densities, including gravitation, and showed that they were conserved as a result of the equations of motion at this order. The existence of these conservation laws also demonstrated, as he expected, that there was no effect of gravitational radiation at 1pN order.

Second post-Newtonian order

After a delay of about four years, during which he worked largely on uniform-density ellipsoids, Chandra finally returned to the problem to develop, in the same systematic way, the much more complicated 2pN order of approximation. He worked on this with his Ph.D. student Yavuz Nutku (Chandrasekhar and Nutku 1969). Again he chose a gauge that he felt was appropriate. Determining that the theory so far still obeyed conservation of energy and angular momentum was necessary but, in view of the complexity of the equations, it was not possible for him to determine the form of the conserved energy and angular momentum simply by inspection. Chandra realized that he could derive the conservation laws systematically from the Landau-Lifshitz pseudotensor, expanded to a suitable order. He justified this in a separate paper (Chandrasekhar 1969), where he showed that this method gave the same results at 1pN order as he had obtained by inspection. In the 2pN paper with Nutku, the same method led to the 2pN-order conservation laws. Again, there was no radiation to this order.

The 2pN equations are of course very unwieldy, and Chandra never used them except as a way of showing that there was nothing left out on the road to radiation reaction order. Nevertheless, Chandra's insight into the key role of the Landau-Lifshitz tensor at this order was to prove crucial to his obtaining the right results for radiation reaction. However, this step did give rise to considerable difficulties for other relativists who wanted to follow Chandra on this road. This is because Chandra's formal expressions for some of the metric terms at this order were divergent.

These divergences came about because Chandra chose a gauge in which, presumably, he felt he could make the equations look as similar as possible to the Newtonian equations. In this gauge, metric corrections played the role of higher-order gravitational potentials, satisfying certain Poisson-like differential equations. The derivatives (with respect to time and space variables) of these potentials contributed to the forces in the equations of motion. These potentials incorporated in part the nonlinearity of the theory, so the sources in their Poisson equations involved, for example, higher-order corrections to the mass density, including the mass associated with the gravitational potential energy of the system. At the 1pN level of approximation, the gravitational potential energy can be localized in the volume containing the perfect fluid, for example in a term like $\rho\Phi/c^2$. But at the 2pN level, the sources for some of the potentials are distributed over all space. While most of them fall off rapidly near infinity, some do not, with the result that the solutions for some of these potentials are formally infinite. This infinity, which Chandra does not comment on in his papers, occurring at an order before the one at which we find radiation reaction, naturally cast doubt on the validity of Chandra's approximation. I shall describe below the way later work cleared up the doubts.

Radiation reaction order

Immediately after finishing the 2pN work, Chandra turned to the equations at radiation-reaction order, $2\frac{1}{2}$pN order, working with another Ph.D. student, Paul Esposito (Chandrasekhar and Esposito 1970). Here the final equations are simpler, since the messy conservative forces of the field appear at even orders. But the derivation of these equations is delicate, because one must apply an outgoing-radiation condition on equations that are a near-zone approximation to the full solution, and this is where much of the previous work on the problem had gone wrong. Chandra read the previous literature extensively, and was attracted to the work of Trautman (1958), who had shown that, if one cast the field equations in the form of wave equations, wrote down solutions using retarded Green functions, and did a near-zone expansion of the result, then one would have a near-zone solution that incorporated the information that the solution was a retarded one and that the gravitational waves (which were not explicitly calculated in this approach) would be outgoing.

Trautman had, nevertheless, not obtained the correct answer, and after some thought Chandra knew why: Trautman had used, as his source for the solutions, only the stress-energy of the fluid system, and Chandra knew that he had to replace this with the full Landau-Lifshitz pseudotensor. This pseudotensor contained the gravitational stress-energy as well, and Chandra understood that this was essential to getting the correct result. The equivalence principle tells us that general relativity treats all energies in the same way, and this means that the gravitational radiation from a system will depend on the lowest-order Newtonian gravitational potential energy (resident in the Landau-Lifshitz pseudo-energy) as much as on, say, the kinetic energy of the fluid source.[7]

In order to incorporate the effects of retardation, Chandra had to alter his method substantially at this order; he chose the de Donder gauge in order to get wave equations, and he had to include in his near-zone expansion new terms from the expansion of the retardation parts of the solutions for the metric. He points out that he is free to adopt the de Donder condition at this order while keeping the previous gauge choices at lower order, so in his method, as contrasted with other approaches, he really uses wave equations for the first time only at this order. This is an unusual approach, not adopted by anyone else, but it is not inconsistent. In this way, Chandra and Esposito calculated the metric and equations of motion by close analogy with Trautman's attempt, but with the Landau-Lifshitz source.

This approach, while physically reasonable, raised another technical problem, but this one does not seem to have led to much criticism. Chandra

---

[7] Others before Chandra, including Landau and Lifshitz themselves and also Fock, had understood that one must include the gravitational stresses to get the correct radiation at infinity, but they had not attempted the near-zone problem of radiation reaction.

found—for the same reason as at 2pN order, namely, the extended nature of the source for his Poisson integrals—that there was one Poisson equation for a metric term where the source was (in the near zone) a constant in space. The solution of such an equation again diverges at infinity. Unlike the divergent terms at 2pN order, Chandra worried about this point and felt he had to address it.

Chandra devised an intriguing method to get around this problem: he required that his metric solution be defined only in what was effectively a distributional sense. That is, he required that the metric give a finite integral when multiplied by any C^∞ function that fell off near infinity faster than any inverse power of r. Using this, he found an expression for the metric that was well defined, albeit divergent as $r \longrightarrow \infty$. However, and importantly, he did not need this metric outside the fluid bodies. It was used to calculate the radiation-reaction force on the system, where this metric term was always multiplied by the mass density of the fluid itself, which was a function in the C^∞ class defined by Chandra.

The radiation reaction result

This is where Chandra reached his goal, a goal that had eluded relativists for decades: the radiation-reaction force in a self-gravitating system. In the context of the time at which Chandra performed this calculation, it seems to me that his key insight was to combine Trautman's boundary condition with the Landau-Lifshitz pseudotensor. This was natural to him because he had labored carefully over the lower orders, where the Landau-Lifshitz complex provided the conserved energy and angular momentum. Chandra's self-described "conservative" approach to the calculation had paid off: he had learned important lessons from the lower-order calculations.

Of course, Chandra knew by the time he derived the reaction force that Burke and Thorne had already found an expression for it by the method of matched asymptotic expansions (Burke 1970; Thorne 1969). Chandra's expression was not identical to the force derived by Burke and Thorne, but it did give the same energy dissipation. This is another matter that Chandra does not comment on in his papers. However, there is a gauge transformation that makes it identical to the simpler Burke-Thorne result, which I quote here. The effects of radiation reaction at this order can be completely described by incorporating into the equations of motion a correction to the Newtonian potential of the form

$$\Phi_{\text{react}} = -\frac{1}{5}\frac{d^5 I_{jk}}{dt^5} x^j x^k \tag{8.1}$$

(Misner *et al.* 1973), where the symbol I_{jk} stands for the reduced or trace-free quadrupole tensor of the Newtonian mass distribution,

$$ \mathcal{I}_{jk} = I_{jk} - \frac{1}{3}\delta_{jk}I_{nn}, \quad I_{jk} = \int \rho x_j x_k d^3 x. \tag{8.2} $$

By changing gauge one can find other forms for this that are equivalent, and some have proved useful in other problems.

8.3   Influence, controversy, and a reassessment

The completion of Chandra's post-Newtonian radiation-reaction work initially attracted attention because it gave, by a completely independent method, a result equivalent to that of Burke and Thorne. This gave astrophysicists confidence that general relativity was physically reasonable and well-behaved, that energy and angular momentum radiated in gravitational waves was correctly balanced by a loss from the energy and angular momentum of the radiating system. There were at least two immediate applications: Chandra's own discovery of the completely unexpected radiation-induced nonaxisymmetric instability in rotating stars (described by John Friedman elsewhere in this volume), and the realization that cataclysmic binary systems can be regulated by the competing effects of the radiation of angular momentum in gravitational waves (which brings the stars closer together) and the transfer of angular momentum from one star to another by mass flows (which in these systems drives the two stars apart; Faulkner 1971).

There was a further direct influence of Chandra's post-Newtonian work, which was the development of the Parametrized Post-Newtonian (PPN) framework for describing a wide class of relativistic theories. The development of space physics and radio astronomy offered new opportunities for testing relativity, and it was initially not clear what the meaning of an experimental result would be if it contradicted a prediction of general relativity. It was clear that any reasonable theory of gravity must reduce to Newton's theory in the appropriate limit. Therefore the first deviations that would distinguish one theory from another would arise at the post-Newtonian level. Theorists developed a framework for describing a large family of relativistic theories of gravity *parameterizing* their post-Newtonian predictions.

The first effort in this direction was by Kenneth Nordtvedt (1967), who in 1968 developed a PPN framework based on the Einstein-Infeld-Hoffman (EIH) point-particle approach. This was generalised to include fluids by Thorne and Will, who were directly motivated by what Thorne had learned about Chandra's post-Newtonian approach during a year-long visit to Chicago. Will (1981) used Chandra's perfect-fluid post-Newtonian equations as the basis for his PPN framework, developed as his Ph.D.

work for Thorne.[8] The PPN formalism of Will, Thorne, and Nordvedt has become the framework for describing the results of tests of general relativity, initially for solar system experiments, but later also for the important Hulse-Taylor binary pulsar, which came shortly after the PPN framework was developed (Will 1981).

The Hulse-Taylor binary pulsar, announced in 1974 (Hulse and Taylor 1975), was soon recognized as potentially the most important application of post-Newtonian theory: it was quickly realised that post-Newtonian effects on the orbits of the two stars would be observable through radiation-reaction order, and that the system would be a test of the correctness of general relativity's description of gravitational radiation. This proved to be so successful that the discoverers were awarded the Nobel Prize in Physics in 1993.

The prospect of testing post-Newtonian theory against the Hulse-Taylor system raised two kinds of questions about Chandra's post-Newtonian work. The first was whether it applied to the orbital motion of the two neutron stars in the Hulse-Taylor binary system even though it clearly did not apply to their internal structure: neutron stars do not have the weak internal fields assumed in Chandra's work. This question could only be resolved by methods that can treat the weak orbital fields without making assumptions about the internal fields. Many workers developed such methods (Futamase 1983; Futamase and Schutz 1985; Damour 1987; Will 1994; Blanchet 1996), and the result was basically that the orbits and interactions of the stars are independent of the compactness of the stars, apart from obvious effects like tidal distortions of one star by the other, and in this system such effects are ignorable.

The second kind of question was more serious for Chandra's work, which was whether it did in fact represent a valid approximation to general relativity, even for systems that were uniformly weak-field. This became an issue because of the divergent terms in Chandra's equations that we described above.

Many relativists, including Thorne (1990) and Chandra himself, did not find these infinities worrying, because they never directly entered his equations of motion. The potentials entered only in their derivatives, and Chandra knew that the derivatives of the integrals were in fact finite. Later work by Kerlick (1980), working on a suggestion by Jürgen Ehlers, showed, by carefully tracking these terms, that the derivatives originally arose inside the integrals, and if they were left inside, then the contributions of these integrals to the equations of motion at 2pN order would be finite, as Chandra himself realized. With simple modifications, therefore, Kerlick

---

[8] Will writes, in a private communication, "In fact my introduction to Chandra's post-Newtonian work was in a term paper I did for [Thorne]'s general relativity course, in which I derived the EIH equations of motion by taking a suitable limit of Chandra's post-Newtonian fluid equations of motion."

showed that Chandra's equations can be made manifestly finite to beyond $2\frac{1}{2}$pN order (Ehlers 1976).

Chandra's method does not, however, continue to indefinitely high order, even with derivatives left inside the integrals. Fundamentally, once radiation is present in the equations, it is impossible to place the equations of motion in an action-at-a-distance, Newtonian-like form. The forces acting on fluid elements at one time are transmitted no faster than the speed of light across the fluid. The appearance of radiation reaction in the equations is the signal that retardation effects can no longer be neglected. It is impossible to construct potentials depending on the state of the fluid at a single time that fairly represent the gravitational forces at that time. Kerlick showed that the method eventually breaks down and can't be repaired. The approaches to the post-Newtonian approximation that have succeeded in going further are all formulated in terms of retarded integrals rather than Poisson-like Green functions (Will 1994; Blanchet 1996).

This limitation on the order to which the method can be pushed has sometimes also been interpreted as casting doubt on the validity or applicability of Chandra's work. Indeed, it certainly shows that Chandra could not have developed in this way a convergent series approximation to general relativity starting at the Newtonian equations; but it seems unlikely to me that Chandra himself ever sought such a series. Chandra wanted calculational tools, the first few terms in a series that could be used to approximate weak-field systems with some accuracy. In this, he was using the (finite number of) terms that he calculated in the post-Newtonian series as an asymptotic approximation to general relativity in the limit of weak fields and slow motion.

This, it seems to me, is a justifiable claim. Approximations based on a finite number of terms can be asymptotic even when further terms in the series get unboundedly large, or indeed when one does not even have a prescription for going to further terms. There is no full proof yet that any post-Newtonian approximation is asymptotic (but see Futamase 1983 for an attempt at a proof, using restrictive assumptions), but it seems plausible in the cases of interest, and it is certainly the working assumption of all the active workers in the field. I believe that one can also make that claim with the same level of confidence for Chandra's post-Newtonian work, with the infinities regularized as described above, and for systems that have uniformly weak gravitational fields (not the Hulse-Taylor system).

Chandra himself remarked to me on more than one occasion that he felt unhappy about the controversy over the divergent terms, because it had prevented him from being given adequate credit in the world of relativity for the significance of his post-Newtonian work. I believe he may have drawn a parallel between his efforts to convince the world of astrophysics about the mass limit on white dwarfs, working against the prejudice of Eddington, and his foray into radiation reaction, where he encountered criticism from many established relativists. In both cases, he was a kind of outsider producing an important result and finding the reception cooler than he expected.

I do not believe that this parallel is very accurate, because Chandra was already a major scientific figure by the time he worked on the post-Newtonian approximations, his work commanding immediate attention and respect; and because the criticisms he encountered from relativists focused on real mathematical shortcomings in his work: they were not simple prejudice against an upstart outsider.

Nevertheless, I agree with him that his work has so far been undervalued. Chandra was the first person to show how, at least conceptually, the radiation-reaction problem could be solved for continuous systems: he put together all the necessary ingredients, the retarded potentials, the Landau-Lifshitz pseudotensor, the near-zone expansion. He had to skate on thin ice over mathematical problems, but these turned out to be largely technical. It is easy today to forget that, at the time, there was considerable confusion in the field. One could find support in published calculations for almost any kind of radiation-reaction formula that one wanted, from the Burke-Thorne-Chandrasekhar-Esposito formula to zero reaction and even to anti-damping.[9] It was, of course, precisely because of this widespread confusion, and the uncertainty about what were the reasons for such wide disagreement, that many relativists wanted to be mathematically careful, and were unhappy with the divergent integrals in Chandra's method.

Chandra did not understand this, at least partly because he had worked through the equations in such detail that he felt he had a physical feeling for the correctness of the methods he used. His was the physicist's intuition that the result was clearly (to him) correct, and if the mathematics was not fully rigorous, that was something that others could clear up. Chandra was more interested in getting to the applications, and his prompt discovery of the radiation-driven instability is itself ample proof that his time was better spent doing new physics than clearing up problems of mathematical rigor!

Chandra was a newcomer to relativity, but by applying his systematic, conservative, exhaustive methods, he became the first to provide a full description of the near-zone, post-Newtonian fluid equations through radiation-reaction order. The newcomer had accomplished, by brute force and close attention to detail, what many relativists over several decades had tried but failed to do.

---

[9] Indeed, this was true even in some papers published ten years later!

8.4   Wider influence: the resurgence of relativity in the 1960s

There is a wider context in which the significance of Chandra's post-Newtonian work should be considered. Chandra played a key role in the movement to bring general relativity into the mainstream of physics and astrophysics. From the perspective of 1996, where general relativity plays a central role in astrophysics, where the goal of the next generation of high-energy physics theories is to unify general relativity with the other forces of physics, and where a course in general relativity is widely regarded as an essential part of a theoretical physicist's training, it may be hard to appreciate how peripheral general relativity was to physics in the 1950s. Its mathematics was unfamiliar to most physicists, even theoretical ones, and the range of physical phenomena to which it seemed relevant was small and unexciting to the leading physicists, who were trying to explain what went on in nuclear reactors and particle accelerators. Most importantly, few graduate students in physics could get any training in general relativity. If you wanted to study the subject, you went to one of a handful of departments that specialized in it. If you studied physics almost anywhere else, you ignored general relativity. In the 1950s, most ambitious physics graduate students would have shared Chandra's perception of relativity as a "graveyard" for their careers.

When changes came, many of them driven by new discoveries in astronomy, relativity was made popular, not by a group of radical Young Turks—as has happened in some other branches of physics—but by established physicists. The blossoming of general relativity seems to have been to a great extent initially a "top-down" phenomenon, in which several well-established physicists with broad backgrounds—among them Hermann Bondi, Bryce DeWitt, Pascal Jordan, Erwin Schrödinger, John Wheeler, and Yakov Zel'dovich—founded schools of research in relativity. They encouraged young scientists to study the subject and helped to ensure that they got their first jobs in the field. This new generation—scientists such as Bruno Bertotti, Jürgen Ehlers, James Hartle, Stephen Hawking, Igor Novikov, Roger Penrose, Dennis Sciama, Kip Thorne, James York, and many others—flourished and showed how important and relevant the field was to astronomy and to the rest of physics. I doubt if these younger people could have begun to study the subject seriously without the protection and encouragement of the established scientists.

Chandra was one of the senior figures, although he began to work in relativity later than the others I have mentioned. His move into relativity, at a time when he was the managing editor of the *Astrophysical Journal*, not only lent legitimacy to the subject but also provided a prestigious journal in which the younger generation, at least in the United States, could publish articles in the respectable new field of relativistic astrophysics. (All the important early papers in the Thorne-Will PPN approach appeared there, for example.) As happened so often in his career, Chandra saw that the time was right to take up a new subject; he played an important role in establishing the subject; and he obtained some of the most fundamental results. Unlike other subjects he worked on, he did not leave relativity after helping to make it important, not even after writing an important monograph (Chandrasekhar 1983). Relativity was his consuming scientific passion for the last 30 years of his life, and he was excited by new problems and new insights in it right up to the time of his death.

8.5   Acknowledgements

I am greatly indebted to the following for helpful conversations and for supplying me with written material that I drew heavily on: Jürgen Ehlers, John Friedman, Norman Lebovitz, Bernd Schmidt, Kip Thorne, and Clifford Will. My conclusions and interpretations are, of course, entirely my own.

References

Anderson, J. L., DeCanio, T. C. 1975, *Gen. Rel. Gravit.*, **6**, 197.

Blanchet, L. 1996, "Gravitational radiation from relativistic sources," in *Astrophysical Sources of Gravitational Radiation*, ed. J.-A. Marck and J.-P. Lasota (Cambridge: Cambridge University Press).

Burke, W. L. 1969, "The coupling of gravitational radiation to nonrelativistic sources," Ph.D. thesis (California Institute of Technology).

Chandrasekhar, S. n.d., unpublished autobiographical memoirs (kindly supplied by N. Lebovitz).

Chandrasekhar, S. 1964, "The dynamical instability of gaseous masses approaching the Schwarzschild limit in general relativity," *Astrophys. J.*, **140**, 417–433.

Chandrasekhar, S. 1965, "The post-Newtonian equations of hydrodynamics in general relativity," *Astrophys. J.*, **142**, 1488–1512.

Chandrasekhar, S., Nutku, Y. 1969, "The second post-Newtonian equations of hydrodynamics in general relativity," *Astrophys. J.*, **158**, 55–79.

Chandrasekhar, S. 1969, *Ellipsoidal Figures of Equilibrium* (New Haven: Yale University Press).

Chandrasekhar, S. 1969, "Conservation laws in general relativity and in the post-Newtonian approximations," *Astrophys. J.*, **158**, 45–54.

Chandrasekhar, S., Esposito, F. P. 1970, "The $2\frac{1}{2}$-post-Newtonian equations of hydrodynamics and radiation reaction in general relativity," *Astrophys. J.*, **160**, 153–179.

Chandrasekhar, S. 1983, *The Mathematical Theory of Black Holes* (Oxford: Oxford University Press).

Damour, T. 1983, "Gravitational radiation and the motion of compact bodies," in *Gravitational Radiation*, ed. N. Deruelle and T. Piran, (Amsterdam: North-Holland), 59–144.

Damour, T. 1987, "The problem of motion in Newtonian and Einsteinian gravity," in *300 Years of Gravitation*, ed. S. W. Hawking and W. Israel (Cambridge: Cambridge University Press), 128–198.

Dirac, P. A. M. 1938, *Proc. Roy. Soc. London*, **A167**, 148.

Ehlers, J., Rosenblum, A., Goldberg, J. N., Havas, P. 1976, *Astrophys. J.*, **208**, L77.

Ehlers, J. 1980, "Isolated systems in general relativity," *Ann. N.Y. Acad. Sci.*, **336**, 279–293.

Faulkner, J. 1971, *Astrophys. J.*, **170**, L99.

Fock, V. A. 1959, *Spacetime and Gravitation* (New York: Pergamon).

Futamase, T. 1983, *Phys. Rev.*, **D28**, 2372–2381.

Futamase, T., Schutz, B. F. 1983, "The Newtonian and post-Newtonian approximations are asymptotic to general relativity," *Phys. Rev.*, **D28**, 2363–2237.

Futamase, T., Schutz, B. F. 1985, "Gravitational radiation and the validity of the far-zone quadrupole formula in the Newtonian limit of general relativity," *Phys. Rev.*, **D32**, 2557–2565.

Hulse, R. A., Taylor, J. H. 1975, "Discovery of a pulsar in a binary system," *Astrophys. J.*, **195**, L51–L53.

Jackson, J. D. 1962, *Classical Electrodynamics* (New York: Wiley).

Kennefick, D. 1996, "Controversies in the history of the radiation reaction problem in general relativity" (preprint).

Kerlick, G. D. 1980, *Gen. Rel. Gravit.*, **12**, 467 and 521.

Landsberg, P. T. 1982, ed., *The Enigma of Time* (Bristol: Hilger).

Misner, C. W., Thorne, K. S., Wheeler, J. L. 1973, *Gravitation* (San Francisco: Freeman).

Nordtvedt, K. Jr. 1968, "Equivalence principle for massive bodies II. Theory," *Phys. Rev.*, **169**, 1017–1025.

Peres, A. 1960, "Gravitational radiation," *Nuovo Cim.*, **15**, 351.

Peters, P. C. 1964, *Phys. Rev.*, **136**, 1224.

Schutz, B. F. 1980, "Statistical formulation of gravitational radiation reaction," *Phys. Rev.*, **D22**, 249–259.

Thorne, K. S. 1969, *Astrophys. J.*, **158**, 1.

Thorne, K. S. 1990, foreword to S. Chandrasekhar, *Collected Papers*, vol. 5, *Relativistic Astrophysics* (Chicago: University of Chicago Press), x–xx.

Trautman, A. 1958, *Bull. Acad. Polon. Sci.*, **6**, 627.

Walker, M., Will, C. M. 1980, *Astrophys. J.*, **242**, L129.

Walker, M. 1984, "The quadrupole approximation to gravitational radiation," in *General Relativity and Gravitation*, ed. B. Bertotti (Dordrecht: Reidel), 109–123.

Will, C. M. 1981, *Theory and Experiment in Gravitational Physics* (Cambridge: Cambridge University Press).

Will, C. M. 1994, "Gravitational waves from inspiralling compact binaries: A post-Newtonian approach", in *Relativistic Cosmology*, ed. M. Sasaki (Tokyo: Universal Academy Press), 83–98.

9

Stability Theory of Relativistic Stars

John L. Friedman

In a personal summary of his work in relativity (Chandrasekhar 1970d), Chandra recalls his early enthusiasm for Eddington's 1931 lectures on the mathematical theory of relativity; but he writes that "the principal reason for [my] reluctance to get seriously interested in relativity was the hardly veiled contempt, I could sense, which physicists like Bohr and others had for the work of Eddington (on fundamental theory) and Milne (on kinematical relativity)."

In 1960, although there was no clear additional evidence that general relativity was soon to play a major role in astronomy, Chandra began to study the field. He did not work in relativistic astrophysics until 1963; and by then, Schmidt's discovery that quasars were at cosmological distances allowed him to write, "The existence of the Schwarzschild limit has been the subject of much recent discussion in the context of astronomical discoveries pertaining to the 'quasistellar' radio sources" (Chandrasekhar 1964a).

9.1 Axisymmetric stability

It was already clear that the great luminosity of a quasar emerged from a compact object of enormous mass. Supermassive stars were proposed as a model, but Chandra's 1964 papers reported a general-relativistic instability to radial oscillations that would essentially rule the model out, implying that such stars were unstable to collapse.

From a relativistic standpoint, the instability can be regarded as a generalization to relativistic gravity of the Chandrasekhar limit on the mass of a white dwarf, because the upper mass limit coincides with the point of instability to collapse. Nonrotating white dwarfs form a 1-parameter sequence of increasing density. At the configuration with maximum mass along this sequence, the fundamental radial mode has zero frequency, because the change from one equilibrium configuration to another with the same mass and larger density is a time-independent radial perturbation. At densities above the maximum mass, the star is unstable.

This connection between maximum mass and instability point holds only for stars whose pulsations and equilibrium are governed by the same effective equation of state. The exact stability criterion that Chandra obtained, however, was more general. It locates an instability point where the second order change in a star's mass vanishes for an adiabatic radial perturbation of a star with arbitrary equation of state.

A Newtonian star is unstable to such perturbations when its average adiabatic index γ is less than 4/3. General relativity's stronger gravity leads to instability at larger values of γ, and it is this fact that implies the instability of supermassive stars, for which the dominance of radiation pressure yields $\gamma \approx 4/3$.

This near equality also holds for dense dwarfs, and it implies that general relativity can render stars dynamically unstable, when they are nearly Newtonian, with radius

$$R = K \frac{2M}{\gamma - 4/3}, \qquad (9.1)$$

where K is a constant of order unity (Chandrasekhar 1964b; Chandrasekhar and Tooper 1964).

Led to the problem by an earlier, heuristic paper by Iben (1963), Chandra saw that "it would be quite straightforward to develop the analog of Eddington's pulsation theory in the exact framework of general relativity." The first paper was completed and corrected "just in time: Misner, Zapolsky and Fowler were already on the trail" (Chandrasekhar 1970a). The previous year, at a lecture by W. Fowler on a supermassive-star model for quasars, Feynman suggested that general relativity might imply instability, because the general-relativistic binding energy increases more rapidly with density than does its Newtonian counterpart, and Fowler followed up on the comment with a post-Newtonian calculation that confirmed Feynman's intuition (Fowler 1964; this account of Feynman and Fowler is taken from Thorne 1990).

Oppenheimer and Volkoff (1939) and Harrison, Wakano, and Wheeler (1958) had already considered the stability of neutron stars using a turning point criterion; shortly after Chandra's paper, Misner and Zapolsky (1964) found numerically that the onset of dynamical instability occurred at configurations of extremal mass. It was soon understood that this coincidence reflected the fact that, in modeling the pulsations, one was using the same effective equation of state as that used to model the equilibrium stars. Thus, as Thorne (1967) subsequently emphasized, the turning-point instability at the maximum mass is technically not dynamical. For masses slightly above the maximum, collapse apparently occurs on a timescale set by the nuclear reactions and energy loss needed to keep the contracting matter in its zero-temperature thermodynamic equilibrium state. (Neutron-star matter is effectively at zero temperature when kT is much smaller than its Fermi energy, roughly $10^{12}°$K). [See Thorne (1967, 1978) for later references and a review of the turning point method applied to spherical stars; a somewhat different treatment is given by Zel'dovich and Novikov (1971)].

The turning-point argument is valid for uniformly rotating stars as well (Friedman, Ipser, and Sorkin 1987) and can be stated heuristically as follows. When the mass has a maximum along a curve of constant J, the total baryon number turns over as well, because of the relation

$$dM = \Omega dJ + \mu dN \qquad (9.2)$$

(Bardeen 1972). At the turning point, nearby models have (to first order in the path parameter ϵ) the same angular momentum, baryon number, and mass. The corresponding perturbation relating two such equilibria is then a time-independent solution to the linearized equations of a perfect fluid in general relativity, but a solution for which the angular momentum of each fluid element changes.

Models on the *high*-density side of the instability point are unstable because the injection energy is a *decreasing* function of central density. The relation can be understood from equation (9.2) if one considers again a sequence of stars with fixed angular momenta. The turning point is a star with maximum mass and baryon number, and on opposite sides of the turning point are corresponding models with the same baryon number. Because $\mu = \partial M/\partial N$ is a decreasing function of central density, the model on the high-density side of the turning point has greater mass than the corresponding model with smaller central density. For spherical neutron stars, the low-density endpoint of the equilibrium sequence is again an extremum of the mass, in this case a minimum. For rapidly rotating stars, however, only the high-density endpoint of a constant J sequence is an extremum of the mass. As the density is lowered at fixed J, the binding energy decreases and the sequence terminates by mass shedding: the equator rotates with angular velocity equal to that of a particle in Keplerian orbit.

The result has a precise phrasing that reflects the J-N symmetry of equation (9.2):

> **Theorem.** Consider a two-dimensional family of uniformly rotating stellar models based on an equation of state of the form $p = p(\epsilon)$. Suppose that along a continuous sequence of models labeled by a parameter λ, there is a point λ_0 at which both $\dot{N} = dN/d\lambda$ and $\dot{J}$ vanish and where $(\dot{\Omega}\dot{J} + \mu\dot{N})^{\cdot} \neq 0$. Then the part of the sequence satisfying $\dot{\Omega}\dot{J} + \mu\dot{N} > 0$ is unstable for λ near λ_0 (Friedman *et al.* 1987).

Cook, Shapiro, and Teukolsky (1994a,b) emphasize the implication that the instability points are extrema of J at constant N, as well as extrema of N and M at constant J (see fig. 9.1).

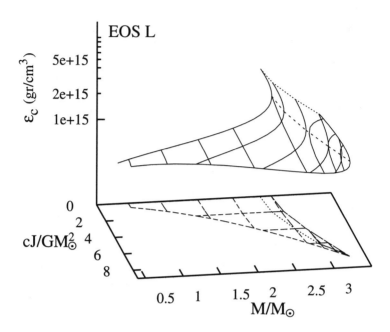

Figure 9.1: The two-dimensional surface of equilibrium models is shown for EOS L. The axes measure angular momentum J, central density ϵ_c, and mass M. The surface is bounded by the spherical $(J = 0)$ and Keplerian $(\Omega = \Omega_K)$ limits and is ruled by solid lines of constant J and constant rest mass M_0. Also shown are the axisymmetric instability sequence (short-dashed line), the projections on the J-M plane (long-dashed lines), and the overlapping of the surface in the J-M plane (dotted lines).

The theoretical expectation of black holes arises from Chandrasekhar's radial instability to collapse. The strongest observational argument in their favor derives from the associated upper limit on the mass of dense spherical stars. Because the equation of state (EOS) is still poorly known above nuclear density, the argument rests on the upper mass limit for the stiffest EOS consistent with causality and with a known EOS below some matching density ϵ_m. For spherical stars, the limit is

$$M < 4.8 M_\odot \left(\frac{2 \times 10^{14} \text{g cm}^{-3}}{\epsilon_m} \right)^{1/2} \tag{9.3}$$

(Hartle and Sabbadini 1977; Rhoades and Ruffini 1974 use a particular choice of ϵ_m). The analogous relation for uniformly rotating models is

$$M < 6.1 M_\odot \left(\frac{2 \times 10^{14} \text{g cm}^{-3}}{\epsilon_m} \right)^{1/2}. \qquad (9.4)$$

(Friedman and Ipser 1987; Koranda *et al.* 1996).

The stabilizing effect of rotation is intuitively clear. A star supported by both rotation and pressure is less dense and has smaller gravity than the corresponding spherical star. A direct study of rotation on the fundamental mode of relativistic stars was considered by Chandrasekhar and Friedman (1971, 1972a, 1972b) following a quasistatic analysis for slowly rotating stars by Hartle, Thorne, and Chitre (1972).

However, this approach relies on nondegenerate perturbation theory, and for isentropic stars at the instability point, the radial mode is degenerate with the set of zero-frequency convective modes. A revised static-stability criterion (Hartle 1975) overcomes the difficulty; it requires the iterative construction of a "comparison sequence" of differentially rotating equilibria, and explicit calculations have been done only for $n = 3/2$ polytropes (Hartle and Munn 1975).

Fortunately, the turning-point method locates the relevant stability points of neutron stars. As in the case of spherical stars, the onset of axisymmetric instability located by the method is initially secular; for rotating stars, its timescale is long enough to accommodate not only heat transfer, but the viscous transfer of angular momentum needed to keep the rotation uniform.

9.2 Nonaxisymmetric instability

Newtonian stars that rotate sufficiently rapidly are unstable to a bar mode, a nonaxisymmetric instability associated with a perturbation having angular dependence $\cos m\phi$, for $m = 2$. Models with viscosity are unstable sooner (at slower rotation) than are perfect-fluid models, because lower energy states with the same total mass and angular momentum are accessible only to perturbations that allow a transfer of angular momentum between fluid rings. The growth time of this secular instability is roughly the time required for viscosity to redistribute angular momentum.

For the uniform density, uniformly rotating Maclaurin spheroids, the instability occurs at a bifurcation point, where the Jacobi family of triaxial ellipsoids branches off (see Chandrasekhar 1969). These ellipsoids are static in a rotating frame, and so is the mode that becomes unstable: at the point of bifurcation it takes the Maclaurin spheroid to a nearby Jacobi ellipsoid. In a conversation with Chandra in 1969, Ostriker raised the question "Does the dissipation of energy by gravitational radiation induce, in the manner of viscosity, a secular instability of the Maclaurin spheroid at the point of bifurcation with the Jacobi sequence?" (Chandrasekhar 1970a). Chandra found that it does not. The mode made unstable by viscosity remains stable when one includes radiation reaction, but there are two surprises that reverse the meaning of this result. Two weeks after completing a paper that reported stability of the Jacobi mode, Chandra rewrote it. He showed that the sequence *is* unstable after the bifurcation point: the instability sets in not by a mode that is static in the rotating frame, but by one that is stationary in the inertial frame (Chandrasekhar 1970b, 1970c).

Bernard Schutz and I subsequently found the second surprise. A nonaxisymmetric instability driven by gravitational radiation is a generic feature of rotating perfect-fluid stars in general relativity (Friedman and Schutz 1978; Friedman 1978; Comins 1979). Every rotating, self-gravitating perfect fluid is unstable to nonaxisymmetric perturbations which radiate away its angular momentum, and the instability first sets in not through the $m = 2$ mode, but through modes of large m. (This is less dramatic than it sounds. As discussed below, viscosity eliminates the instability in ordinary stars and sharply limits its role even in neutron stars).

The nonaxisymmetric instability arises in the following way. For slowly rotating stars, gravitational waves remove positive angular momentum from the forward-moving mode and negative angular momentum from the backward-moving mode, and they therefore damp all oscillations. Once the angular velocity of the star is sufficiently large, however, a mode that moves backward relative to the star is dragged forward relative to an inertial observer. Gravitational radiation will then remove positive angular momentum from the mode. But a mode that moves backward relative to the fluid has negative angular momentum, because the perturbed fluid does not rotate as fast as it did without the perturbation. Gravitational radiation thus removes positive angular momentum from a mode whose angular momentum is negative. By making the angular momentum of the perturbation increasingly negative, gravitational radiation drives the mode.

For a nonrotating star, modes with angular dependence $e^{i(\sigma t \pm m\phi)}$ are degenerate. If, in a rotating frame, slow rotation changes σ only slightly, then in an inertial frame, the forward mode $(e^{-im\phi})$ will be sped up and the backward-moving mode $(e^{+im\phi})$ slowed down. That is, if we denote the ϕ-coordinates of inertial and rotating observers by ϕ_I and ϕ_R, respectively, we have $\phi_I = \phi_R + \Omega t$, and the phase of the mode has the form

$$\sigma t \pm m\phi_I = \sigma_R t \pm m\phi_R, \tag{9.5}$$

where the frequency σ measured in the inertial frame is given, in terms of the frequency σ_R measured by an observer on the star, by

$$\sigma = \sigma_R \pm m\Omega. \tag{9.6}$$

Although σ_R itself depends on Ω, the degeneracy is split in the way suggested by this equation. A mode becomes unstable when its frequency σ

vanishes. From equation (9.4), it appears that for any angular velocity Ω, modes with m greater than some critical value on the order of Ω/σ_0 will be unstable, where σ_0 is the frequency of the unperturbed mode. This conclusion is correct.

Dynamical stability of a star is governed by the sign of the energy of its perturbations. (The gravitational-wave-driven instability is dynamical in the sense used here, because the dynamical behavior of the gravitational field cannot be separated from the fluid dynamics in the exact theory.)

Associated with the time-translation symmetry vector t^a of a stationary star is a conserved current j^a describing the energy flow of its perturbations. Although the current is gauge-dependent, the energy

$$E = \int_S dS_a j^a, \qquad (9.7)$$

obtained by integrating it over an asymptotically null or spacelike hypersurface, is not.

Along a family S_t of asymptotically spacelike hypersurfaces, E is conserved. Along a family S_t of asymptotically null hypersurfaces, however, it changes in time due to the radiation of energy at future null infinity. This radiated energy may again be expressed as a surface integral of j^a, this time at null infinity. Because the surface integral at null infinity is positive, E decreases monotonically from one asymptotically null hypersurface S_t to another S_t' in its future.

The functional E was first constructed in the spherical case by Chandra in his 1964 papers. In my thesis work with Chandra, the functional was generalized to axisymmetric configurations (Chandrasekhar and Friedman 1972a,b, 1973a), and Schutz (1972) simultaneously obtained an equivalent functional. Chandra and I found a related criterion for nonaxisymmetric stability by noting that if one formally exchanged the role of t and ϕ, a time-dependent axisymmetric perturbation became a time-independent nonaxisymmetric perturbation (1973b). With Lebovitz, Chandra obtained the Newtonian limit of this criterion (Chandrasekhar and Lebovitz 1973), and Chandra himself generalized it to differentially rotating configurations (1974).

This criterion is not a minimum principle, however, while the energy functional is; and E was obtained for nonaxisymmetric perturbations (Friedman and Schutz 1975; Friedman 1978) by using the more covariant formalism of Taub (1969) and Carter (1973).

The first papers on the generic nonaxisymmetric instability mentioned only in passing its possible damping by viscosity. Results of a study of Detweiler and Lindblom (1977) suggested that viscosity would stabilize any mode whose growth time was longer than the viscous damping time, and this was confirmed by Lindblom and Hiscock (1983). Our present understanding of the viscosity of neutron stars is summarized by Lindblom and Mendell (1995) and in a more detailed review of the structure and stability of rotating relativistic stars (Friedman and Ipser 1992). (See also Lai and Shapiro (1995) for a reevaluation of bulk viscosity, first discussed by Sawyer [1989a,b].)

At present, it appears that the gravitational-wave-driven instability can limit neutron-star rotation only in hot stars, with temperatures above the superfluid transition point. For substantially higher temperatures, bulk viscosity may also damp the instability. If neutron stars with weak magnetic fields are formed in the accretion-induced collapse of white dwarfs, their rotation may be limited by this instability. If our understanding of viscosity is roughly correct, however, old neutron stars spun up by accretion will never be hot enough to be unstable to gravitational-wave-driven modes. (As noted below, if the equation of state is stiff enough, old, accreting neutron stars with large masses may be unstable to a viscosity-driven bar mode.)

The location of the instability points in the exact theory has been found numerically for the first time (Stergioulas 1996; Stergioulas & Friedman 1996). A giant code written by Stergioulas finds instability points for the $m = 2$, 3 and 4 modes of polytropes. These occur at values of the angular velocity that are smaller than expected from the Newtonian studies of Managan (1985) and of Imamura *et al.* (1985). The code is based on the two-potential formalism of Ipser and Lindblom (1991a,b) (see also Yoshida and Eriguchi 1995; Lindblom 1995).

A striking feature of this work is a discovery that the $m = 2$ (bar) mode can become unstable for much softer polytropes than is the case in the Newtonian theory. Uniformly rotating Newtonian polytropes are stable to the $m = 2$ mode unless $n < 0.8$ ($\gamma = 1 + 1/n > 2.2$) (James 1964). For models near the maximum mass, however, relativistic polytropes can exhibit an $m = 2$ instability for n as large as 1.5 ($\gamma > 1.7$).

The destabilizing effect of general relativity that Chandra found for the fundamental radial mode is present for the nonaxisymmetric modes as well. For an $n = 1$ polytrope, modes with $m = 3$–5 are unstable for dimensionless values of Ω that are substantially smaller than in the Newtonian limit (see fig. 9.2). More striking is the fact that the $m = 2$ mode, which in uniformly rotating stars is unstable only for very stiff EOS ($n < 0.8$), is unstable in the most relativistic stars for n as large as 1.5.

Chandra thought that the nonaxisymmetric instability would play a significant part in gravitational collapse. Because of the high numerical viscosity of codes that model the evolution of fluids, one cannot yet use them to investigate the gravitational-wave-driven instability. A recent Lai-Shapiro study (1995) avoids the 3+1 numerical problem by examining the instability in the collapse of classical ellipsoids, within a post-Newtonian framework, following the earlier work of Miller (1974) (done while she was a graduate student with Chandra) and of Detweiler and Lindblom (1977;

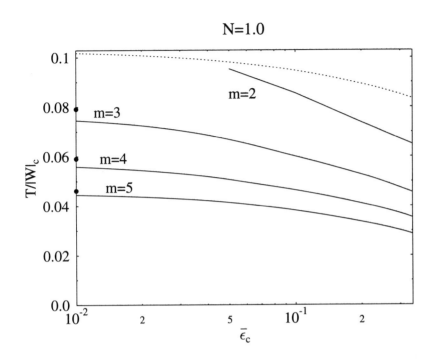

Figure 9.2: Critical ratio of rotational to gravitational binding energy versus a dimensionless central energy density $\bar{\epsilon}_c$ for the $m = 2$, 3, 4, and 5 neutral modes of $n = 1.0$ polytropes. The largest value of $\bar{\epsilon}_c$ shown corresponds to the most relativistic stable configurations, while the lowest $\bar{\epsilon}_c$ corresponds to less relativistic configurations. The filled circles on the vertical axis represent the Newtonian limit while the dotted line is the Kepler limit.

see also Imamura *et al.* 1995 for a study of an analogous secular instability in which angular momentum is removed from the star not by gravitational waves but by the coupling of the star to an accretion disk). In these studies, gravitational radiation makes the bar mode ($m = 2$) unstable, and it is important in the collapse. Lai and Shapiro emphasize that enough energy is radiated in gravitational waves that, if neutron stars are formed by the accretion-induced collapse of white dwarfs, the collapsing stars may be candidates for gravitational-wave interferometers. Incompressible ellipsoids, however, allow much more rapid rotation, measured by the dimensionless quantity $T/|W|$, than do uniformly rotating neutron stars. A uniformly rotating neutron star reaches its maximum rotation well before the large values of $T/|W|$ considered by Lai and Shapiro. But they argue that with a rotation law that is only slightly differential, $T/|W|$ can be large enough for the bar mode to be important.

The viscosity-driven and dynamical bar instabilities that are a key part of Chandra's work on ellipsoids *are* accessible to numerical evolution, and recent studies (Houser, Centrella, and Smith 1994; Smith *et al.* 1996; Bonnazola *et al.* 1995; Durisen and Tohline 1985; Tohline *et al.* 1985; Durisen *et al.* 1986; Williams and Tohline 1988) show what can be done. For $T/|W| > 0.27$ (much larger than the maximum value for uniformly rotating neutron stars), Newtonian stars are dynamically unstable to a bar mode. The work on dynamical instability by Houser, Centrella, and Smith and the earlier authors, examines collapse with substantial differential rotation, and the evolution they trace shows a bar that grows into spiral arms. As it loses angular momentum and drops below the critical rotation needed for dynamical instability, the spiral arms wrap themselves around a symmetric core, and the configuration returns to axisymmetry. Because the axisymmetric form is still rotating rapidly enough to be secularly unstable ($T/|W| > 0.14$), Lai and Shapiro suggest two stages of nonaxisymmetric instability—a dynamical bar instability during the collapse, and a post-collapse instability in which the core evolves to a nonaxisymmetric configuration on a secular timescale. The possibility should also be mentioned that a star-disk or core-envelope coupling drives a nonaxisymmetric instability (Imamura *et al.* 1995).

To summarize: A dynamical bar instability is a feature of accretion-induced collapse of white dwarfs in which rotation is rapid enough that $T/|W|$ exceeds 0.27. Whether a post-collapse secular instability arises is less certain. In particular, although the possible formation of rapidly rotating neutron stars from the accretion-induced collapse of white dwarfs was suggested more than 25 years ago, we do not know whether neutron stars form in this way; and we do not know whether the magnetic field of a neutron star formed in this way can be small enough to allow rapid rotation. If the magnetic field does not limit the star's initial rotation, it is likely that gravitational radiation will drive a nonaxisymmetric instability. But our understanding of the dominant mechanisms for effective viscosity, of the equation of state of neutron-star matter, and of the likely amount of differential rotation in the newly-formed star is too primitive to know which mode will dominate or to be certain that the instability driven by gravitational radiation will not be damped by viscosity.

Finally, for neutron stars spun up by accretion, there is some chance for a nonaxisymmetric instability driven by viscosity. (As mentioned earlier, the viscosity in old, cold stars is almost certainly too large to damp the radiation-driven modes; but gravitational radiation reaction is then too small to damp a mode driven by viscosity.) Although the radiation-driven $m = 2$ mode sets in sooner in general relativity than in the Newtonian theory, Bonnazola *et al.* (1995) find, within their approximation scheme, that in relativity, the mode driven by viscosity sets in at larger values of rotation (measured by the dimensionless quantities $T/|W|$ or Ω/Ω_K). An exceptionally stiff equation of state is therefore needed to allow a viscosity-driven instability of a 1.4 $M_\odot$ star, but stars near their upper mass limit may be unstable for mid-range equations of state.

9.3 Epilogue

Osterbrock's account of Chandra and his students does not touch on years he devoted to general relativity, and a brief discussion is appended here. Chandra did not enter relativity wholeheartedly until the late sixties, when he had finished the classical ellipsoid work. He called himself a graduate student in relativity and was proud that it was more natural for him to interact and collaborate with students than with senior colleagues. Until 1971, he was still sole editor of the *Astrophysical Journal*, but Chandra spent as much time on research as did his most dedicated students. Beginning by 5 A.M., he finished each 13-hour workday late in the evening.

By 1968, pulsars had been discovered and were recognized as neutron stars. This, together with Chandra's understanding that his relativistic instability made supermassive stars an implausible model for quasars, must have convinced him that general relativity was to play a major role in astronomy. He accepted a large group of students, with Philip Greenberg, Paul Esposito, Yavuz Nutku, and, later, Roger Stettner working on aspects of the post-Newtonian approximation, and Persides on conservation laws and the asymptotic gravitational field. Most of the rest of us worked in stability theory—Bonnie Miller (1974) and I looked at the stability of rotating stars, Bonanos at the stability of the Taub universe, and Nduka at the Roche problem.

To bring himself and his students up to date in general relativity, Chandra brought a spectacular array of tutors to the University. The year before I came to Chicago, in 1966–7, Roger Penrose had presented the Newman-Penrose spinor-based formalism that later played a key role in understanding the normal modes and stability of black holes. The next year Kip Thorne taught a quarter of relativistic astrophysics. Our own private summer school followed in 1968: with the clarity and elegance of his lectures serving as his argument, Robert Geroch converted us to a covariant viewpoint and a formalism tied to abstract indices that became a signature of Chicago's younger relativists. Geroch and Brandon Carter taught us about black hole uniqueness; in fact Carter had barely finished his uniqueness theorem before giving his lectures. George Ellis replayed for us his influential Cargese lectures on relativistic cosmology.

The next year we were treated to Andrej Trautman, as meticulous as Chandra in his attention to historical scholarship. Trautman taught us and the rest of the physics community that gauge theories were the physicists' rediscovery of E. Cartan's connections on fiber bundles—the generalization for both communities of the electromagnetic vector potential. And the Sciama-Kibble theory of torsion was a similar rediscovery of Cartan's generalization of the metric connection. It was a couple of years before he (and then we) learned that the Hopf fibration was the Dirac monopole, and that both had been discovered in the same year.

By 1976, Chandra began to speak about retiring from science, and he took no more students. Although he spoke about retiring for the next twenty years, he continued to work nearly as hard as ever. And in place of his own graduate students, he recruited those of his colleagues. The remarkable work that computed, for the first time, the outgoing modes of black holes was done in collaboration with Steven Detweiler (then a student of Jim Ipser). In the numerical calculations that have been done of black-hole formation, these modes turn out to dominate the gravitational waves emitted, and their computation is, in my view, the most important part of Chandra's study of black holes. Chandra, however, held in higher regard his extensive mathematical study of black hole perturbations that followed this work, a study in which a collection of miraculous relations emerges from calculations that span hundreds of pages. Closely related to the mathematical theory of black holes was Chandra's examination of colliding-wave geometries, done in collaboration with Basilis Xanthopoulos, who had been a student of Robert Geroch. When Xanthopoulos, among the warmest and most enthusiastic scientists I have known, was later shot and killed by a student, Chandra played an essential role in establishing the Xanthopoulos Prize in relativity. Chandra's final collaboration, with Valeria Ferrari, led to the discovery of a remarkable set of outgoing modes of relativistic stars that have no Newtonian counterpart.

The beautiful hand in which his equations were written mirrored Chandra's understanding of the equations themselves. For most physicists on the mathematical side, equations are viewed abstractly in a way that highlights the properties their expressions share as operators on a Hilbert space, while astrophysicists usually take from mathematics only what is needed for the problem at hand. Chandra, however, fell in neither camp. For his time, Chandra was, to my knowledge, unique in the way he treated the equations of relativistic astrophysics seriously as objects in themselves, their structure clear in the manner he displayed them, their meaning to be found in this structure. That mathematics was the language of nature he never doubted, and he served nature all his life.

Chandra was also unique in the way he combined a deep understanding of classical mathematics, of astrophysics, and of the history of science, particularly the history of classical physics and astronomy. Trautman and Roger Penrose were then the physicists to whom Chandra seemed closest in temperament and perspective, while his interests were those of the astrophysical relativists, Kip Thorne and James Bardeen. The understanding that grew from Chandra's history distinguished the problems he worked on, and the unmatched artistry with which he handled his language of equations distinguished their solutions.

References

Bardeen, J. M. 1972, in *Black Holes*, ed. C. and B. S. DeWitt (New York: Gordon and Breach).

Bonnazola, S., Frieben, J., Gourgoolhon, E. 1995 (to appear).

Carter, B. 1973, *Comm. Math. Phys.*, **30**, 261.

Chandrasekhar, S. 1964a, *Phys. Rev. Lett.*, **12**, 114; erratum, *Phys. Rev. Lett.*, **12**, 437.

Chandrasekhar, S. 1964b, *Ap. J.*, **140**, 417.

Chandrasekhar, S. and Tooper, R. F. 1964c, *Ap. J.*, **139**, 1396.

Chandrasekhar, S. 1969, *Ellipsoidal Figures of Equilibrium* (New Haven: Yale).

Chandrasekhar, S. 1970a, "General relativity" (unpublished).

Chandrasekhar, S. 1970b, *Phys. Rev. Lett.*, **24**, 611.

Chandrasekhar, S. 1970c, *Ap. J.*, **161**, 561.

Chandrasekhar, S., Friedman, J. L. 1971, *Phys. Rev. Lett.*, **12**, L14.

Chandrasekhar, S., Friedman, J. L. 1972a, *Ap. J.*, **175**, 379.

Chandrasekhar, S. and Friedman, J. L. 1972b, *Ap. J.*, **176**, 745.

Chandrasekhar, S. and Friedman, J. L. 1973a, *Ap. J.*, **181**, 481.

Chandrasekhar, S. and Friedman, J. L. 1973b, *Ap. J.*, **185**, 1.

Chandrasekhar, S. and Lebovitz, N. R. 1973, *Ap. J.*, **185**, 19.

Chandrasekhar, S. 1974, *Ap. J.*, **187**, 169.

Comins, N. 1979, *Mon. Not. R. Astr. Soc.*, **189**, 233 and 255.

Cook, G. B., Shapiro, S. L., Teukolsky, S. A. 1994a, *Ap. J.*, **422**, 227.

Cook, G. B., Shapiro, S. L., Teukolsky, S. A. 1994b, *Ap. J.*, **424**, 823.

Detweiler, S. L., Lindblom, L. 1977, *Ap. J.*, **213**, 193.

Durisen, R. H., Tohline, J. E. 1985, in *Protostars and Planets II*, ed. D. Black and M. Matthews (Tucson: University of Arizona Press).

Durisen, R. H., Gingold, R. A., Tohline, J. E., Boss, A. P. 1986, *Astrophys. J.*, **305**, 281.

Fowler, W. A. 1964, *Rev. Mod. Phys.*, **36**, 545 and 1104.

Friedman, J. L. 1978, *Comm. Math. Phys.*, **62**, 247.

Friedman, J. L., Ipser, J. R. 1987 *Ap. J.*, **314**, 594.

Friedman, J. L., Schutz, B. F. 1975, *Ap. J.*, **200**, 204.

Friedman, J. L., Schutz, B. F. 1978, *Ap. J.*, **222**, 281.

Friedman, J. L., Ipser, J. R., Sorkin, R. D. 1987, *Ap. J.* (in press).

Friedman, J. L., Ipser, J. 1992, in *Classical General Relativity* (New York: Oxford University Press).

Harrison, B. K., Wakano, M., Wheeler, J. A. 1958, in *Onzieme Conseil de Physique Solvay, La Structure et l'evolution de l'universe* (Brussels: Stoops).

Hartle, J. B. 1975, *Ap. J.*, **195**, 203.

Hartle, J. B., Munn, M. W. 1975, *Ap. J.*, **198**, 467.

Hartle, J. B., Sabbadini, A. G. 1977, *Ap. J.*, **213**, 831.

Hartle, J. B., Thorne, K. S., Chitre, S. M. 1972, *Ap. J.*, **176**, 177.

Houser, J. L., Centrella, J. M., Smith, S. C. 1994, *Phys. Rev. Lett.*, **72**, 1314.

Iben, I. 1963, *Ap. J.*, **138**, 1090.

Imamura, J. N., Friedman, J. L., Durisen, R. H. 1985, *Ap. J.*, **294**, 474.

Imamura, J. N., Toman, J. Durisen, R. H., Pickett, B. K., Yang, S. 1995, *Ap. J.*, **444**, 363.

Ipser, J. R., Lindblom, L. 1991a, *Ap. J.*, **373**, 213.

Ipser, J. R., Lindblom, L. 1991b, *Ap. J.*, **379**, 285.

James, R. A. 1964, *Astrophys. J.*, **140**, 552.

Koranda, S., Stergioulas, N., Friedman, J. L. 1996 (in preparation).

Lai, D., Shapiro, S. L. 1995, *Ap. J.*, **442**, 259.

Lindblom, L., Mendell, G. 1995, *Ap. J.*, **444**, 804.

Lindblom, L., Hiscock, W. A. 1983, *Ap. J.*, **267**, 384.

Lindblom, L. 1995, *Ap. J.* (in press).

Managan, R. 1985, *Ap. J.*, **294**, 463.

Miller, B. D. 1974, *Ap. J.*, **187**, 609.

Misner, C. W., Zapolsky, H. S. 1964, *Phys. Rev. Lett.*, **12**, 635.

Oppenheimer, J. R., Volkoff, G. M. 1939, *Phys. Rev.*, **55**, 374.

Rhoades, C. E., Ruffini, R. 1974, *Phys. Rev. Lett.*, **32**, 324.

Sawyer, R. F. 1989a, *Phys. Rev.*, **D39**, 3804.

Sawyer, R. F. 1989a, *Phys. Lett.*, **B233**, 412.

Schutz, B. F. 1972, *Ap. J. Suppl.*, **24**, 343.

Smith, S. C., Houser, J. L., Centrella, J. M. 1996, *Ap. J.*, **458**, 236.

Stergioulas, N. 1996, "Structure and stability of rotating relativistic stars," Ph.D. thesis, (Milwaukee: University of Wisconsin).

Stergioulas, N., Friedman, J. L. 1996 (in preparation).

Taub, A. H. 1969, *Comm. Math. Phys.*, **15**, 235.

Thorne, K. S. 1967, in *High Energy Astrophysics*, vol. 3, ed. C. DeWitt, E. Schatzman, and P. Veron (New York: Gordon and Breach).

Thorne, K. S. 1978, in *Theoretical Principles in Astrophysics and Relativity*, ed. N. R. Lebovitz, W. H. Reid, and P. O. Vandervoort (Chicago: University of Chicago Press).

Thorne, K. S. 1990, foreword to S. Chandrasekhar, *Selected Papers*, vol. 5, *Relativistic Astrophysics* (Chicago: University of Chicago).

Tohline, J. E., Durisen, R. H., McCollough, M. 1985, *Ap. J.*, **298**, 220.

Williams, H. A., Tohline, J. E. 1988. *Ap. J.*, **334**, 449.

Yoshida, S., Eriguchi, Y. 1995, *Ap. J.*, **438**, 830.

Zel'dovich, Ya. B., Novikov, I. D. 1971, *Relativistic Astrophysics*, vol. 1 (Chicago: University of Chicago Press).

10

Chandrasekhar, Black Holes, and Singularities

Roger Penrose

10.1 Historical background

The dilemma that was presented to the scientific world by Chandrasekhar's early work (1931) on the existence of a maximum mass for white dwarf stars took some while to be fully appreciated. There were some, such as Eddington, who did seem to understand the alarming implications of Chandra's conclusions. Assuming the correctness of the relativistic equations of state, it seemed that a white dwarf star of mass more than about 1.4 solar masses would have to collapse inwards, its density increasing indefinitely as the body approached a singular configuration at the center. However, Eddington himself regarded this as a *reductio ad absurdum*, concluding instead that there must be something wrong with Chandra's use of the relativistic equations. Eddington supposedly had in mind that some new physical principles must come into play in order to save the star, perhaps such as were embodied in his own approach to a deeper fundamental theory (Eddington 1946). Taken at the level of phenomena at which Chandra's discussion was intended to apply, there is no doubt that Chandra's analysis was the correct one, as the experts seem to have appreciated even at that time (at least privately), despite the weight that Eddington's authority attached to the contrary view. Yet Eddington had a point: the impossibility of an equilibrium state would lead to the star's unstoppable collapse. Would this collapse continue until the star becomes so compressed that it reaches its Schwarzschild radius ($r = 2m$), thought at that time to be a dimension at which the very metric structure of space-time becomes singular? In any case, as the star continues to collapse radially inwards, it appears that it should reach a state where the density becomes infinite. And, according to Einstein's general theory of relativity, infinite matter density would in itself imply a singularity in space-time structure. Thus, it would seem that even if one accepted the procedures of conventional physics completely, with the conclusions that Chandra so strongly argued for, the unending collapse of the star would lead to a situation in which those very procedures would ultimately have to be abandoned.

By nature conservative in his approach, Chandra would certainly not have been attracted to the radical kind of speculation that had so occupied Eddington in his later years. In the 1930s, general relativity was not regarded as a worthy activity for aspiring astrophysicists to devote themselves to. In any event, for whatever reason, Chandra chose not to mount a direct attack on the problem of gravitational collapse. In that regard, he bided his time. Indeed, there were many other issues that would occupy his attentions for a quarter of a century!

Around 1960, having completed his works on stellar structure and stellar dynamics, on radiative transfer, and on the equilibrium or stability of various structures of astrophysical interest, he finally resolved to enter into a study of general relativity. He attended the 1962 International Conference on General Relativity and Gravitation in Warsaw as a "student," in order to attain an overview of current research in the subject. Even after he had thoroughly prepared himself, he did not directly address the issue of the fate of a collapsing star. He was more concerned, first, with the effects of general relativity on the stability of gravitating bodies, and then with general-relativistic corrections to the Newtonian dynamics of collections of masses, including the effects of gravitational radiation damping. These studies were relevant to the issue of how considerations of general relativity might affect the onset of gravitational collapse, not with the result of the collapse itself. In short, the overall conclusion was that the influence of general relativity accelerates gravitational collapse somewhat, over and above the various influences of other physical effects.

So was collapse right down to a singularity of some sort the inevitable conclusion? For a white dwarf star, there might still be a number of other possibilities open to it. It might, for example, have the opportunity to find ultimate rest as a neutron star. But it was already an implication of Chandra's analysis (cf. Landau's simplified approach of 1932) that it applied also to the relativistic equations of state for neutron matter. The analysis was carried out in detail by Oppenheimer and Volkoff (1939), the conclusion at that time being that the maximum mass for a neutron star was a little smaller, even, than for a white dwarf. Allowing for the fact that the neutrons themselves can become converted to other massive particles (not known at the time), somewhat larger limiting masses were subsequently suggested. On the other hand, very general considerations, based on fundamental principles such as causality, allowed the conclusion to be made that there was an absolute overall limit of not much more than 3 solar masses. There appears to be no clear agreement as to what the actual limit is, but observations seem to suggest something like about 1.4 solar masses. Accordingly, no final solution to the problem of gravitational collapse is to be found along these lines.

Already in 1939, Oppenheimer and Snyder had faced this problem squarely, considering the situation of a collapsing spherically symmetrical dust cloud, of uniform density, treated according to Einstein's general relativity. In their description, they provided the first explicit model of a collapse to what is now called a "black hole." They showed that although the so-called "Schwarzschild singularity," which occurred at radius $r = 2m$, is not actually a singularity—what is now called the "horizon"—there is still a space-time singularity at the center, where the density of the dust indeed becomes infinite. Ironically, it was Eddington himself who first published, in 1924, a metric form for the spherically symmetric Schwarzschild space-time in which the "Schwarzschild singularity" is exhibited in its true nature as a null hypersurface (actually two null hypersurfaces). However, Eddington appeared not to appreciate what he had himself done. Before Oppenheimer and Snyder, Lemaître (1933) had appreciated that the "Schwarzschild singularity" could be locally eliminated by a coordinate change, but he had not provided an overall picture of collapse to a black hole. Later Synge (1950) and others found the complete extension of the Schwarzschild solution.

Since there is still a singularity in the Oppenheimer-Snyder collapse model (at $r = 0$), the Chandrasekhar dilemma is not removed by their collapse picture. However, many people remained unconvinced that this description would necessarily be the inevitable result of the collapse of a star too massive to be sustainable as either a white dwarf or a neutron star. There were a number of good reasons for some skepticism. In the first place, the equations of state inside the matter were assumed to be those appropriate for pressureless dust, which is certainly far from realistic for the late stages of stellar collapse. Moreover, the density was assumed to be constant throughout the body. With realistic material, there are many alternative evolutions to that described by Oppenheimer and Snyder. For example, nuclear reactions set off at the center could lead to an explosion— a *supernova*—which might perhaps drive off sufficient mass from the star that a stable equilibrium configuration becomes possible.

Most serious of all was the assumption of exact spherical symmetry. Since, in this picture, all the material of the body is aimed directly at the central point, a resulting density singularity could easily be the result merely of this fortuitous focusing. It could be expected that the introduction of even the slightest perturbation away from spherical symmetry might cause most of the inward falling particles of the body's material to miss the central point, so that even though the density might get very large there, it might well not diverge to infinity.

Indeed, as late as 1963, the Russian school of Lifshitz and Khalatnikov (1963) had claimed that the "generic" solution of the Einstein equations ought to be free of singularities. However, it became evident that there must be a significant error in their work when the early singularity theorems were established (Penrose 1965; Hawking 1966; Hawking and Penrose 1970). In fact, such an error was found and corrected by Belinskii (cf. Belinskii, Khalatnikov, and Lifshitz 1970, 1972), their final conclusion being that singularities could occur in generic solutions after all. (Lifshitz and Khalatnikov had omitted a certain degree of freedom in their expansions.)

In any case, it is difficult to form any firm conclusions about the final stages of gravitational collapse from the kind of power series analysis that the Russian school had adopted. Perturbation analysis is not well suited to the extreme situations that would be expected to arise when a space-time approaches a singular state. Moreover, the possibility of finding explicit solutions of sufficient generality to describe what happens in a realistic collapse could be virtually ruled out. For reasons such as these, the approach adopted in the singularity theorems was completely different. Instead of attempting to work out what happens in detail to a collapsing star, general overall considerations of a largely topological nature were employed so as to derive general properties of the solution. In essence, the conclusions were all of a negative character, in the sense that they ruled out certain things, showing that they cannot happen—rather than showing what actually does happen.

What cannot happen, according to these theorems, is an ordinary singularity-free evolution—assuming that the equations of state satisfy a reasonable energy-positivity condition—if the collapse reaches a certain point of no return. This "point of no return" can be characterized as the existence, in the space-time, of what is called a "trapped surface." This is a compact spacelike 2-surface whose null normals converge into the future (which means that if a flash of light were to be emitted at the surface, then the area of any element of cross-section of the emitted light rays must decrease; cf. Penrose 1965). Another slightly different way of specifying an appropriate "point of no return" is the existence of a point in the space-time through which all light rays into the future begin to reconverge somewhere (cf. Hawking and Penrose 1970). By an "ordinary singularity-free evolution," I mean that it must be possible to continue the space-time (non-singularly) into one in which all null and/or timelike geodesics have infinite affine length and for which appropriate causality conditions are maintained (such as the existence of a global Cauchy hypersurface or merely the absence of closed timelike curves together with a condition that the space-time is in some mild sense "generic"). There are slight differences in the details of these various conditions, depending upon which singularity theorem is being appealed to. (For further information, see Hawking and Penrose 1970; Hawking and Ellis 1973).

The upshot of all this is that once a trapped surface (or reconverging light cone) has formed, there is no way, within the scope of existing physical laws, to extend the space-time indefinitely. This is the ultimate dilemma that Eddington was, in effect, shying away from. But why should we expect a trapped surface to arise in any case? In the Oppenheimer-Snyder picture, just after the collapsing matter shrinks within the Schwarzschild radius at $r = 2m$ (the event horizon), there are trapped surfaces in the space-time region immediately surrounding the body. Since the trapped-surface condition is, by its very nature, something undisturbed by (adequately small) finite perturbations, the essential issue is whether or not a collapsing body or collection of bodies is ever likely to reach the vicinity of its Schwarzschild radius—and just beyond. Chandra's work on the influences of general relativity on the stability of massive bodies, and (even more) his much earlier initiation of the entire line of thinking concerning the maximum mass of bodies held apart by relativistic degeneracy pressure, provided strong support for the view that unstoppable collapse to the neighborhood of the Schwarzschild radius was probable.

Nevertheless, these arguments are not totally convincing, owing to the lack of complete information about details of the internal state of a star, where the density might exceed, by several orders of magnitude, even that inside an atomic nucleus. However, if it is matters of principle that we are concerned with, these details are not important. One can envisage situations in which trapped surfaces will arise even for densities that are as low as, say, that of air. By making the total mass of the collapsing system large enough, the density at which the body crosses its Schwarzschild radius can be made as small as we please. At the centers of large galaxies there would be collections of ordinary stars which, if they were to find themselves in a small enough region all at once, would be surrounded by a trapped surface even though the stars are not yet in contact, so the total density is less than stellar density. This follows from very basic considerations of general relativity—even directly observed ones—concerning the focusing power of mass density on light rays. Rather easier to use than the trapped-surface condition is the condition of a reconverging light cone. Suppose that the radius of the region is 10^4 km. The light rays emerging from a point at the center of the star cluster would encounter sufficient stellar material that they would indeed begin to reconverge, as follows from a simple order-of-magnitude calculation. (See Penrose 1969, p. 266, for a simple description of this idea.)

It follows from this, and the singularity theorems, that if conventional physical ideas hold true, we are forced into having to face up to the occurrence of space-time singularities. We must ask what is the nature of these singularities and what we are to do about them. In fact, the singularity theorems are almost completely silent about the nature of the singularities themselves. The theorems are simply existence theorems, and say almost nothing about the location of the singular regions, let alone anything about their detailed nature. From later work (e.g., Clarke 1993, and also the earlier work of Belinskii, Khalatnikov, and Lifshitz 1970, 1972) there were certain very incomplete indications as to the singular behavior of the Weyl curvature tensor. Enough is known to suggest that the structure of the singularities in a generic collapse will be quite different from that which is encountered in the big bang (divergent Weyl curvature in the former, essentially vanishing Weyl curvature in the latter—a feature intimately connected with the second law of thermodynamics, cf. Penrose 1979). But very little is known in detail. As to "what we are to do about them," it is clear that these singularities take us outside the domain of classical general relativity. Without the appropriate union between general relativity and quantum mechanics being to hand, we are presented with an impasse.

The existence of singularities does not, however, imply the existence of black holes. This deduction requires the additional assumption of what is called "cosmic censorship." Cosmic censorship (Penrose 1969, 1978) asserts that naked singularities will not occur in a generic gravitational collapse. Such a singularity, roughly speaking, would be one that could be seen by an outside observer. Cosmic censorship would imply that the region lying to the future of all singularities resulting from gravitational collapse cannot reach future null infinity $\mathscr{I}^+$. The boundary of the entire region which cannot be connected to $\mathscr{I}^+$ by a causal (i.e., timelike or null) curve defines the (absolute) *event horizon H*. We thus see that a gravitational collapse resulting in singularities subject to cosmic censorship will lead to the existence of a *horizon H* which "shields" all the singularities from directly revealing themselves to the outside world. This horizon describes a black hole (see Penrose 1973).

Various theorems (Israel 1967; Carter 1970; Robinson 1975; Hawking 1972) can now be called into play, where it is assumed that the Einstein vacuum (or Einstein-Maxwell) equations hold from the neighborhood of H out to infinity, after the collapsing body has fallen through. The upshot of these theorems is that if the space-time in this region settles down to a stationary configuration, then it is described by the Kerr metric (Kerr 1963), completely defined by two parameters m and a (or, in the Einstein-Maxwell case, the Kerr-Newman metric (Newman *et al.* 1965), defined by three parameters m, a, and e).

10.2 Chandra's work on perturbations of black holes

It is a remarkable fact that black holes, when they settle down to an exactly stationary configuration, have such a precise and explicit mathematical description. Chandra never ceased to be impressed by this fact, remarking at the beginning of his prologue to *The Mathematical Theory of Black Holes* (1983):

> The black holes of nature are the most perfect macroscopic objects there are in the universe: the only elements in their

construction are our concepts of space and time. And since the general theory of relativity provides only a single unique family of solutions for their descriptions, they are the simplest objects as well.

Chandra's study of black holes was based upon these stationary cases: the Kerr (and Kerr-Newman) space-times—together with the Schwarzschild (and Reissner-Nordström) specializations. Since these stationary metrics were known explicitly, he was able to study, in comprehensive detail, the first-order gravitational perturbations away from the stationary Kerr configuration—and gravitational-electromagnetic perturbations in the Reissner-Nordström case. Moreover, he provided a comprehensive treatment of the Maxwell and Dirac equations on a Kerr background. In all these cases, he found remarkable algebraic/differential relations which enabled him to separate and decouple the equations.

Chandra was a relative latecomer to the study of black holes. In his early work on white dwarf stars and the inevitability of their collapse when too massive, leading to his early realization of the dilemma referred to in section 10.1, Chandra had been far ahead of his time. But his assault upon the very problem that his early researches had thrown up was delayed until after most of the groundwork had been carried through by others. There were, of course, excellent reasons for this. Chandra's monumental work on other topics had to be completed first. It was his way of working that he would devote himself in a single-minded way to one topic at a time, thereby achieving the phenomenal thoroughness and depth that he strove for in each topic.

However, he also had a long-standing interest in general relativity and, specifically, in what came to be known as black holes. But until about 1974 (about the time when most experts in the subject had moved away from the classical theory and were turning to the implications of Hawking's discovery of black-hole radiation), he had not felt ready to embark upon his detailed assault upon the area of black holes.

In particular, he felt that he needed to complete what had been his most recent work—on the stability of rotating stars—before doing so. As it happens, in this work, he was studying perturbative general-relativistic effects, including the effects of gravitational radiation. Thus, this research provided him with a natural route into the study of perturbations of black holes. Indeed, the various researches that Chandra has undertaken should not be thought of as totally independent of one another. In particular, in his rotating star stability work, as in so many other things that he achieved, he exploited his extraordinary ability with equations, and he already had much of the basic framework of ideas to hand, ready for his concerted attack on the problems of black-hole perturbation. Indeed, it appears to have been Teukolsky's separation of the equations for gravitational perturbations of a Kerr black hole (Teukolsky 1973) that stimulated Chandra's actual entry into the subject.

The physical motivations for the study of black hole perturbations came from a desire to know how such an object, initially in a stationary configuration, would react if slightly disturbed. Its response could involve the emission of gravitational waves, with another part of the disturbance disappearing into the hole. It would be argued that this response would be governed primarily by the structure of the black hole itself. The linear perturbations of the hole would be described by linear equations, and could therefore be analyzed in terms of the appropriate "modes." These are not quite like the modes of vibration of a perfectly elastic body, because of the effective dissipation that occurs both through gravitational radiation and through loss into the hole itself. The frequencies are therefore complex, with imaginary parts describing the decay of the modes.

Chandra's first paper on black-hole perturbations (Chandrasekhar 1975) was concerned with the Schwarzschild black hole. At that time there already existed a thorough analysis of these perturbations, dating back to the work of Regge and Wheeler in 1957. In particular, Zerilli (1970) had shown that, splitting the perturbation into different components for the individual spherical harmonics, each component satisfies a Schrödinger equation, corresponding to that for a wave of time dependence $e^{i\sigma t}$ (with complex frequency σ having an imaginary part describing the damping rate) and spatial dependence $Z(x)$, impinging on a potential barrier defined by a certain smooth potential function $V(x)$:

$$\left(\frac{d^2}{dx^2} + \sigma^2\right) Z = VZ.$$

The variable x is related to the standard Schwarzschild radial coordinate r by $x = r + 2m \log(r - 2m)$. The potential function V is a particular explicit function of r, depending on the mass m and the choice of spherical harmonic. (Units are always chosen so that the speed of light c and Newton's gravitational constant G are both unity.) Accordingly, the general perturbation problem for the Schwarzschild black hole can be reduced to that of finding the transmission and reflection coefficients of a simple one-dimensional barrier-penetration problem in quantum mechanics. On the other hand, Bardeen and Press (1973) had obtained a different set of equations to describe the same perturbations, and various relations between all these expressions and procedures had remained somewhat mysterious. Rederiving all these expressions in his own way, Chandra was able to understand and explain several of these mysterious features, most notably a curious relationship between the coefficients that arise from the odd- and even-parity (or, as Chandra preferred, axial and polar) perturbations.

Later, he sought to analyze the deeper reasons underlying such relationships, noticing that one could understand these in terms of the procedures of inverse scattering, according to which consistency conditions of the nature of the Korteweg–de Vries equation arise (Chandrasekhar 1982). In this way, Chandra provided an opening into the intriguing and mathematically fruitful area of integrable systems, and it is likely that the last word on these matters has by no means been heard (see comments in section 10.4).

The initial disturbance which causes a perturbation to the geometry of a stationary black hole could be caused by some physical object falling into the hole, or it could be taken to have the form of incoming gravitational (or gravitational-electromagnetic) waves impinging on the hole from outside. Particularly in his later work on the subject, Chandra preferred to emphasize the latter viewpoint, this having the advantage that no foreign ingredients are imported into the system of equations under consideration, everything being described in terms of the Einstein vacuum equations (or Einstein-Maxwell equations). Of course, in an actual astrophysical situation, there would normally be some other source for a significant black-hole perturbation, but from the point of view of making the mathematical treatment self-contained, this approach has considerable advantages.

In a situation of this nature, there is an incoming component (from $\mathscr{I}^-$) and two "outgoing" components, one which escapes out to infinity ($\mathscr{I}^+$) and another one which falls into the hole. As mentioned above, the situation is closely analogous to that which arises with one-dimensional potential scattering and barrier penetration in ordinary quantum mechanics—and for each separate spherical harmonic, the mathematical description is precisely of this form. There is an incoming wave train and an outgoing reflected wave train, accompanied by the part which passes through the potential barrier. In particular, there will be that situation which arises when the incoming influence is "switched off" and the black hole "rings" according to its natural frequencies. These modes are what are called "quasi-normal modes," characterized by the fact that there is no component coming in from $\mathscr{I}^-$ (and no component coming out from the interior of the hole—a geometrical impossibility in any case unless one allows for a delayed burst of radiation from the collapsing matter which originally produced the hole, which would have to have hovered, exponentially decaying, at the hole's horizon since its formation). Thus, each quasi-normal mode is composed only of a wave train escaping to $\mathscr{I}^+$ and a wave train falling into the hole. Each of these would be exponentially decaying modes (but badly behaved at spatial infinity owing to their exponential blow-up at infinite negative times). Chandra studied these modes for the Schwarzschild black hole in his 1975 paper with Detweiler (Chandrasekhar 1975) (and, later, for the Kerr black hole), but he warned against too much reliance on these giving a complete characterization of the decay behavior of black-hole perturbations and on too much faith being placed on the analogy with normal modes of an elastic material body. As far as I am aware, there are still unanswered questions concerning completeness and other issues for quasi-normal modes for black holes.

After his comprehensive treatment of the spherically symmetrical (i.e., Schwarzschild) case, Chandra moved on to his study of perturbations of the Kerr space-times. These possess axial symmetry and include angular momentum. It was, after all, the remarkable simplicity of general relativity's implication that stationary vacuum black holes have to be Kerr (or Schwarzschild) metrics that so attracted him to this area of research. The fact that Teukolsky (1973) was able to separate the radial parameter r and the angular coordinate θ to obtain a pair of decoupled equations for the gravitational perturbations of the Kerr metric was a surprising additional bonus. Chandra rederived the Kerr metric in his own way (Chandrasekhar 1978a). With Detweiler (Chandrasekhar and Detweiler 1975b), he showed that Teukolsky's equations for the gravitational perturbations can be reduced to a (complex) one-dimensional wave equation of the type considered in the displayed equation above, with four possible potentials (but now depending on the frequency σ); moreover, they showed that the reflection and transmission coefficients are the same in each of the four cases, thereby illuminating some puzzling relationships. This work was continued in an important series of papers (Chandrasekhar 1976a,b, 1978a,b,c, 1979a,b, 1980, 1983) in which Chandra completed a very thorough and detailed analysis of the gravitational perturbations of a Kerr black hole. These papers are full of remarkable algebraic and differential relationships and identities, following from those that had been established by Teukolsky and Starobinski.

Teukolsky (1973) had also separated the equations for scalar waves and electromagnetic fields on a Kerr background. Chandra treated both of these fields in two papers in 1976, again reducing the equations to the same type of one-dimensional wave equation as above. He then showed (Chandrasekhar 1976c), ingeniously performing separation prior to decoupling, that the Dirac equation for the electron could also be separated and decoupled in a Kerr background. Teukolsky's work had already covered the case of a neutrino field

$$\nabla^{AB'} \nu_{B'} = 0,$$

but for a Dirac field with nonzero mass $2^{1/2}\mu$, in 2-component spinor form, we have the coupled pair

$$\nabla_{AA'} P^A + i\mu \, \bar{Q}_{A'} = 0,$$

$$\nabla_{AA'} Q^A + i\mu \, \bar{P}_{A'} = 0,$$

and the separation of these had proved to be a stumbling block. Chandra took special delight in the fact, as noted in half a sentence at the end of his paper, that by taking the limit in which the black hole's mass is set to zero, one obtains the separation of Dirac's equation in oblate spherical coordinates in flat space-time—a feat which had not been achieved before.

In his study of the Dirac equation, Chandra showed that he had mastered the 2-spinor formalism, which is not particularly familiar to physicists generally, there being an almost universal tendency for them to phrase their discussions of the Dirac equation in terms of the (superficially simpler but ultimately more complicated) 4-spinor formalism. In this, and also in much of his work in gravitational and electromagnetic perturbations (and in his later work on colliding plane waves), Chandra exhibited a great facility with what has become known as the "NP-formalism" (the method of spin coefficients), which Ted Newman and I had developed in 1962 to handle general relativity—effectively by combining the 2-spinor calculus with a Ricci-rotation-coefficient type of formalism (Newman and Penrose 1962). Chandra had specifically invited me to give a series of lectures in Chicago, in the early 1970s, primarily on the subject of this formalism, but I had in no way anticipated the powerful use that he would ultimately make of it.

To this, I might add a personal note of some irony. Some time later, in response to a question from me as to why he had not gone further and adopted the somewhat more streamlined later GHP formalism (Geroch, Held, and Penrose 1973), Chandra had remarked that this would not simplify his equations in the way that I had imagined, because of an awkward problem about normalizing the spinor dyads against one another. Consequently, when writing the book *Spinors and Space-Time* with Wolfgang Rindler, I went to some trouble to develop a generalized version of the GHP formalism in which the normalization condition was removed. This resulted in some extra complications, which I know caused certain of my colleagues some irritation. However, I was not deterred by this, partly because the original GHP formalism can be extracted from this without difficulty, but more particularly because I believed that this was all in an excellent cause because Chandra could then directly incorporate this extended formalism into simplifying his equations! Apparently Chandra did not agree that this was any help, so we are left with a formalism still (as far as I know) looking for a good application. (Actually, I am still not convinced that it cannot be used with effect in the kind of thing that Chandra was doing. Perhaps some brave soul will have a look at it sometime.)

Chandra's work in relation to the Kerr metric applied to a background space-time in which the Einstein vacuum equations hold. Returning to the case of spherical symmetry, but where now the presence of an electromagnetic field is allowed for, Chandra studied gravitational-electromagnetic perturbations away from a stationary Reissner-Nordström black hole (Chandrasekhar 1979b; Chandrasekhar and Xanthopoulos 1979). Again, separation and decoupling of the perturbations occurs, with apparently rather little—but sometimes subtle—change required from the vacuum Schwarzschild case.

There is, however, one important change which does take place when one passes from the Schwarzschild to the Reissner-Nordström black hole, a change which occurs also when one passes to the Kerr black hole, this being the acquisition of a Cauchy horizon. A space traveler who falls into the hole would, upon crossing the inner (Cauchy) horizon, enter a "new universe," were it not for the fact that a shell of infinitely intense radiation coming from outside the hole would be expected to be encountered there. Some early work (cf. Simpson and Penrose 1973; McNamara 1978a,b) had provided a strong indication of this, but a more complete analysis of the perturbation theory, provided by Chandra with James Hartle in 1982, convincingly demonstrated the inevitability of this phenomenon (Chandrasekhar and Hartle 1982).

The combination of angular momentum with electric charge in a black hole gives rise to the solution of the Einstein-Maxwell equations known as the Kerr-Newman metric (Newman *et al.* 1965). Chandra also studied Kerr-Newman black holes and their perturbations, but he was disappointed that he was unable to separate the equations for the gravitational perturbations, and he finally set the problem aside. If Chandra was not able to do it, that in itself would seem to be reason enough to believe that separation is not actually possible. However, the analogies with the Kerr and Reissner-Nordström black holes are very strong, so it is hard to resist the temptation to suspect that some separation, perhaps of a much more complicated nature, might lie somewhere behind the scenes. (In connection with this, following Chandra's separation of the Dirac equation on a Kerr background, Page was able, in 1976, to extend this result to the Kerr-Newman background.) I shall return briefly to the question of separation in section 10.4.

As was his general method of working, Chandra virtually set the seal on nearly ten years of research into this topic—black holes and their perturbations—by writing a superb book on the subject: *The Mathematical Theory of Black Holes*, published by the Oxford University Press in 1983. Unlike the situation with his earlier books, however, he did not leave black holes entirely alone after that (cf. Chandrasekhar 1984, 1989; Chandrasekhar and Xanthopoulos 1989; Chandrasekhar 1990; Chandrasekhar and Ferrari 1990). Moreover, he certainly did not leave the subject of general relativity aside, as we shall see.

10.3   Colliding plane waves

It appears that the initial impetus that led to the next stage of Chandra's work, namely that on colliding plane waves in general relativity, was a letter from Yavuz Nutku (who had many years earlier been one of his students) which pointed out that the metric that arises when two impulsive gravitational waves with non-parallel polarization collide is described by the simplest solution of certain equations that Chandra had encountered in his derivation of the Kerr metric (Chandrasekhar 1978a).

There seems to be little doubt that it was Chandra's fascination with the mathematics of black holes that gradually began to turn him away from a rigorous requirement that his work be directly relevant to astrophysically realistic situations. Of course, his study of black-hole perturbations was indeed astrophysically relevant, particularly because gravitational wave detectors may, before too long, possibly be able to detect the "ringing" of a black hole after its formation or after it swallows a companion body. However, as his work continued, he became more and more seduced by the remarkable *mathematical* quality of the equations that he encountered.

Perhaps his 1989 paper with Xanthopoulos on two black holes attached to strings is the most extreme—and least "Chandra-like"—of his publications in its departure from astrophysical realism. But colliding impulsive plane waves are also somewhat unrealistic. Since a plane wave must extend all the way to infinity, one cannot expect such a wave to be realized accurately in the physical universe. Moreover, impulsive gravitational waves (where the curvature tensor has the form of a Dirac delta function) do not provide a very reasonable idealization of the distant gravitational field of a violent event (such as, say, the congealing of two black holes) owing to the fact that there is an unending constant flux of gravitational radiation energy after the impulsive wave has passed. In addition, the nature of the space-time singularity that arises after the encounter between two such impulsive waves may have a special and unrepresentative structure, owing to its arising from a situation in which there is exact symmetry and precise focusing.

All this notwithstanding, colliding waves may well have provided an ideal framework for Chandra to make substantial progress towards an understanding of the very problem that his early researches into the equilibrium of white dwarfs had led to: the existence of *space-time singularities*. Colliding plane waves indeed have the habit of leading to such singularities; and there is at least the possibility that these singularities may be closer to being realistic than those which occur in the Schwarzschild black hole (too special symmetry) and the Kerr black hole (closed timelike curves and intervening Cauchy horizon). It is not clear to me that Chandra was much concerned by the issue of physical realism at this stage in any case. Here was a family of solutions that he could study in detail. He could bring all of his magical gifts with equations to bear on these examples and, with luck, some deeper understanding of the nature and formation of space-time singularities could indeed come about.

His first paper on this subject described work that he and Valeria Ferrari had done (Chandrasekhar and Ferrari 1984) concerning the very situation that Nutku had described in his letter (Nutku and Halil 1977) [and which generalized the situation that Khan and I published in 1971, where the planes of polarization of the approaching impulsive plane waves were taken to be parallel (Khan and Penrose 1971)]. Their paper provided a very comprehensive analysis of this (Nutku-Halil) space-time, including detailed expressions for all the NP spin-coefficients and curvature quantities.

In his next three papers on the subject, this time joint work with Xanthopoulos (Chandrasekhar and Xanthopoulos 1985a,b, 1986a) he included various matter terms in the equations: electromagnetic, perfect fluid (with $\varepsilon = p$), and null dust. Again the treatment was very detailed, but there were certain complicating issues that arose. For example, it was not reasonable to admit a delta function in the electromagnetic component to the incoming plane waves. For a delta function in the electromagnetic field would lead to a square of a delta function in the energy-momentum tensor and therefore in the Ricci curvature, which is not allowable. For there to be a delta function in the Ricci tensor, to accompany the already present delta function in the Weyl tensor describing the gravitational impulse, the Maxwell tensor would have to have the form of a square root of a delta function. Chandra circumvented this particular problem by having the Maxwell field have only a step function at the gravitational impulse.

One source of potential confusion arose from the unusual way in which Chandra tended to obtain his solutions to the various field equations for this situation. Figure 10.1 (taken from Chandrasekhar and Xanthopoulos 1986a) represents a space-time diagram for those two coordinates for which there is dynamic evolution. These are taken to be two null coordinates u and v (whose sum and difference may be regarded as the "time" and the "distance in the direction of propagation"), while the remaining two coordinates describe the flat plane surfaces of the waves—these planes being orbits of the two commuting translational symmetries of the space-time. Region IV is the flat space-time between the approaching impulsive waves. Regions II and III are the portions of the two waves to the far side of the leading impulse. Region I is where the scattering between the waves takes place, this being the only region where serious work needs to be carried out in solving equations.

The flat portion of the space-time, prior to the arrival of either wave, is region IV. Chandra's ease with equations made it natural for him to start with the scattering region, and then to see what kind of waves in regions II and III might give rise to whatever solution for region I he might have found! This may seem strange to those physicists who think in terms of the evolution from Cauchy data on an initial hypersurface. In Chandra's case of the perfect fluid with $\varepsilon = p$ (the case illustrated in the fig. 10.1), it turned out that in order to obtain the required solution for region I, the material in regions II and III could be taken to be null dust. The apparent difference between the equations of state holding in region I and those holding in regions II and III caused some confusion and a certain

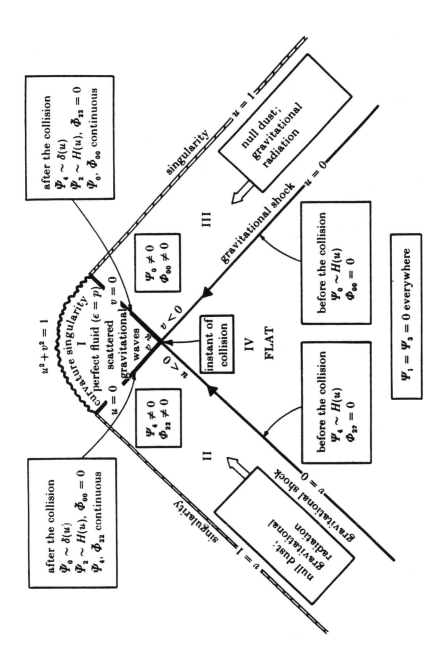

Figure 10.1: Colliding plane waves.

amount of controversy. Basically this was, I believe, a misunderstanding of Chandra's position concerning this situation. In fact, the perfect fluid with $\varepsilon = p$ has null dust as a limiting configuration, so there is not really any inconsistency between the different regions.

To clarify this point, the energy momentum tensor for an $\varepsilon = p$ fluid has the form

$$T^{ab} = 2v^a v^b - g^{ab} \left(g_{cd} v^c v^d \right),$$

where the vector v^a is directed along the fluid's 4-velocity and satisfies

$$v^a v_a = \varepsilon = p.$$

When v^a becomes a null vector, we get $\varepsilon = p = 0$, but this is consistent, and the energy-momentum tensor becomes $T^{ab} = 2v^a v^b$, namely, null dust. Thus, null dust can be regarded as a particular (limiting) state of an $\varepsilon = p$ fluid.

In three subsequent papers (Chandrasekhar and Xanthopoulos 1986b, 1987a,b), Chandra confronted the issue of the singularity itself and the curiously slippery nature of the development of singularities in general relativity. The first of these papers presented a result which was a genuine surprise to me when Chandra first informed me about it. I had been under the impression (though without proof) that the circular line marked "curvature singularity" at the top of figure 10.1 would always remain a region of infinite curvature for all collisions of this type. However, Chandra told me of his work in progress with Xanthopoulos on a vacuum colliding wave space-time, providing an example which, at that stage, could be seen to have nondiverging curvature there. Soon after discussing the matter with me, Chandra was able to see how to change his coordinates so as to continue the metric explicitly across that seemingly singular region, exhibiting it to have more subtle geometric character than I had imagined, with a null hypersurface appearing that plays a role similar to that of an event horizon, and beyond which lies a timelike singularity. The remaining two coordinates cannot be ignored in this discussion, and the structure of the entire space-time remains somewhat complicated. In the other two papers, analogous situations are considered in which an electromagnetic field or an $\varepsilon = p$ fluid is introduced. The curiously contrasting behaviors are examined in some detail. Chandra's work with Xanthopoulos was clearly a high point for him, and it was a particular tragedy when Xanthopoulos met his untimely death from the attack of a crazed assassin. Chandra wrote a moving tribute in his *Selected Papers*, vol. 6 (the volume that Xanthopoulos edited). Later he played an important role in the establishment of the triennial Xanthopoulos prize, for work in relativity achieved by young researchers.

10.4   Aspects of Chandra's mathematical heritage

The papers that I have discussed above are far from exhausting Chandra's voluminous research into general relativity. There are several other articles in which various aspects of exact solutions are discussed. It is striking how his involvement with general relativity seduced him more and more in the direction of mathematics, where his earlier requirements of direct relevance to astrophysics seem to have been somewhat pushed aside. It seems clear that the equations that he encountered in his work on general relativity theory gave him immense satisfaction, and he seemed continually surprised at the mysterious beauty that these equations revealed. In his book *Truth and Beauty: Aesthetics and Motivations in Science* (Chandrasekhar 1987), he remarked on this explicitly; and in his booklet entitled *The Series Paintings of Claude Monet and the Landscape of General Relativity*, his dedication lecture on the opening of the Centre for Astronomy and Astrophysics in Pune (Chandrasekhar 1992), he compared some of this mathematical beauty with a sequence of paintings by Monet. Particular instances of the vacuum equations that had impressed him so much had occurred in a comparison that he had found between the basic black-hole space-times (Schwarzschild, Kerr) and the corresponding basic colliding wave space-times (Khan-Penrose, Nutku-Halil). A similar comparison was pointed out for the Einstein-Maxwell equations. All this arose from the mathematical structure of solutions of the Ernst equation (Ernst 1968), which controls the solutions of Einstein vacuum (or Einstein-Maxwell) metrics with two commuting symmetries.

As far as I am aware, these comparisons have not yet been fully understood in terms of the various techniques that have been developed over the years to handle such space-times. The particular procedure that appeals to me the most is that developed by Woodhouse and Mason (1988), which uses the procedures of twistor theory to describe these space-times in terms of holomorphic vector bundles over non-Hausdorff Riemann spaces. This is part of a general comprehensive treatment of integrable systems (of which space-time metrics with two commuting Killing vectors provide an example) in terms of twistor theory, which these two authors have thoroughly developed in a recent book (Mason and Woodhouse 1996). In view of the many intriguing relationships with different features of integral systems that Chandra's work has thrown up, I feel sure that there is a good deal that is deep, yet to be learned, from a study of the insights that he gained from his work in general relativity.

The same can also be said of a study of his analysis of the separation of gravitational perturbations and of other systems of equations in stationary black-hole backgrounds. There is yet much mystery to be unraveled. Some of this has already been achieved in the work of Carter (1968), Walker and Penrose (1970), Carter and McLenaghan (1979), Kamran and McLenaghan

(1984), and many others, whereby separation can be related to the existence of a Killing tensor, Killing spinor, and Killing-Yano tensor. There are relations to twistor theory here also, and it is my guess that a further study of Chandra's work from this direction may well throw some profound light on these issues.

In a sense, Chandra's lifetime work was like a circle, starting with his insights that led us to believe that excessively massive white dwarfs must collapse to space-time singularities, and finally reaching back to a sophisticated study of those very singularities. Yet, for all Chandra's extraordinary ingenuity and industry, the deep answers are still missing. I rather believe that Chandra did not really expect that this work might directly find an answer, however. The main thrust of his work in this area was somewhat different. It was driven more and more by the quest for mathematical elegance, coupled with a deep belief in a profound underlying connection between physics and mathematics. Perhaps it is here that the answers are finally to be found.

References

Bardeen, J. M., Press, W. H. 1973, *J. Math. Phys.*, **14**, 7.

Belinskii, V. A., Khalatnikov, I. M., Lifshitz, E. M. 1970, *Adv. Phys.*, **19**, 523.

Belinskii, V. A., Khalatnikov, I. M., Lifshitz, E. M. 1972, *Soviet Phys. JETP*, **62**, 1606.

Carter, B. 1968 "Hamilton–Jacobi and Schrödinger separable solutions of Einstein's equations," *Commun. Math. Phys.*, **10**, 280.

Carter, B. 1970. "An axisymmetric black hole has only two degrees of freedom," *Phys. Rev. Lett.*, **26**, 331.

Carter, B., McLenaghan, R. G. 1979, "Generalized total angular momentum operator for the Dirac equation in curved space-time," *Phys. Rev.*, **D19**, 1093.

Chandrasekhar, S. 1931, "The maximum mass of ideal white dwarfs," *Astrophys. J.*, **74**, 81.

Chandrasekhar, S. 1975, "On the equations governing the perturbations of the Schwarzschild black hole," *Proc. R. Soc. London*, **A343**, 289.

Chandrasekhar, S., Detweiler, S. 1975a, "The quasi-normal modes of the Schwarzschild black hole," *Proc. R. Soc. London*, **A344**, 441.

Chandrasekhar, S., Detweiler, S. 1975b, "On the equations governing the axisymmetric perturbations of the Kerr black hole," *Proc. R. Soc. London*, **A345**, 145.

Chandrasekhar, S. 1976a, "On a transformation of Teukolsky's equation and the electromagnetic perturbations of the Kerr black hole," *Proc. R. Soc. London*, **A348**, 39.

Chandrasekhar, S. 1976b, "The solution of Maxwell's equations in Kerr geometry," *Proc. R. Soc. London*, **A349**, 1.

Chandrasekhar, S. 1976c, "The solution of Dirac's equation in Kerr geometry," *Proc. R. Soc. London*, **A349**, 571.

Chandrasekhar, S. 1978a, "The Kerr metric and stationary axisymmetric gravitational fields," *Proc. R. Soc. London*, **A358**, 405.

Chandrasekhar, S. 1978b, "The gravitational perturbations of the Kerr black hole. I." *Proc. R. Soc. London*, **A358**, 421.

Chandrasekhar, S. 1978c, "The gravitational perturbations of the Kerr black hole. II." *Proc. R. Soc. London*, **A358**, 441.

Chandrasekhar, S. 1979a, "The gravitational perturbations of the Kerr black hole. III." *Proc. R. Soc. London*, **A365**, 425.

Chandrasekhar, S. 1979b, "On the equations governing the perturbations of the Reissner-Nordström black hole," *Proc. R. Soc. London*, **A365**, 453.

Chandrasekhar, S., Xanthopoulos, B. C. 1979, "On the metric perturbations of the Reissner-Nordström black hole," *Proc. R. Soc. London*, **A367**, 1.

Chandrasekhar, S. 1980, "The gravitational perturbations of the Kerr black hole. IV. The completion of the solution," *Proc. R. Soc. London*, **A372**, 475.

Chandrasekhar, S. 1982, "On the potential barriers surrounding the Schwarzschild black hole," in *Spacetime and Geometry: The Alfred Schild Lectures*, ed. R. A. Matzner and L. C. Shepley (Austin: University of Texas Press).

Chandrasekhar, S., Hartle, J. B. 1982, "On crossing the Cauchy horizon of a Reissner-Nordström black hole," *Proc. R. Soc. London*, **A384**, 301.

Chandrasekhar, S. 1983, *The Mathematical Theory of Black Holes* (Oxford: Clarendon Press).

Chandrasekhar, S. 1984, "On algebraically special perturbations of black holes," *Proc. R. Soc. London*, **A392**, 1.

Chandrasekhar, S., Ferrari, V. 1984, "On the Nutku-Halil solution for colliding impulsive gravitational waves," *Proc. R. Soc. London*, **A396**, 55.

Chandrasekhar, S., Xanthopoulos, B. C. 1985a, "On colliding waves in the Einstein-Maxwell theory," *Proc. R. Soc. London*, **A398**, 223.

Chandrasekhar, S., Xanthopoulos, B. C. 1985b, "On the collision of impulsive gravitational waves when coupled with fluid motions," *Proc. R. Soc. London*, **A402**, 37.

Chandrasekhar, S., Xanthopoulos, B. C. 1986a, "On the collision of impulsive gravitational waves when coupled with null dust," *Proc. R. Soc. London*, **A403**, 189.

Chandrasekhar, S., Xanthopoulos, B. C. 1986b, "A new type of singularity created by colliding gravitational waves," *Proc. R. Soc. London*, **A408**, 175.

Chandrasekhar, S. 1987, *Truth and Beauty: Aesthetics and Motivations in Science* (Chicago: University of Chicago Press).

Chandrasekhar, S., Xanthopoulos, B. C. 1987a, "On colliding waves that develop time-like singularities," *Proc. R. Soc. London*, **A410**, 311.

Chandrasekhar, S., Xanthopoulos, B. C. 1987b, "The effect of sources on horizons that may develop when plane gravitational waves collide," *Proc. R. Soc. London*, **A414**, 1.

Chandrasekhar, S. 1989, "The two-centre problem in general relativity: The scattering of radiation by two extreme Reissner-Nordström black holes," *Proc. R. Soc. London*, **A421**, 227.

Chandrasekhar, S., Xanthopoulos B. C. 1989, "Two black holes attached to strings," *Proc. R. Soc. London*, **A423**, 387.

Chandrasekhar, S. 1990, "The Teukolsky-Starobinsky constant for arbitrary spin," *Proc. R. Soc.*, **A430**, 433.

Chandrasekhar, S., Ferrari, V. 1990, "The flux integral for axisymmetric perturbations of static space-times," *Proc. R. Soc.*, **A428**, 325.

Chandrasekhar, S. 1992, *The Series Paintings of Claude Monet and the Landscape of General Relativity* (Pune: Inter-University Center for Astronomy and Astrophysics).

Clarke, C. J. S. 1993, *The Analysis of Space-Time Singularities*. Cambridge Lecture Notes in Physics (Cambridge: Cambridge University Press).

Eddington, A. S. 1924, "A comparison of Whitehead's and Einstein's formulas," *Nature*, **113**, 192.

Eddington, A. S. 1946, *Fundamental Theory* (Cambridge: Cambridge University Press).

Ernst, F. J. 1968, *Phys. Rev.*, **168**, 1415–1417.

Geroch, R., Held, A., Penrose, R. 1973, "A space-time calculus based on pairs of null directions," *J. Math. Phys.*, **14**, 874–881.

Hawking, S. W. 1966, "The occurrence of singularities in cosmology," *Proc. R. Soc. London*, **A294**, 511.

Hawking, S. W. 1972, "Black holes in general relativity," *Comm. Math. Phys.*, **25**, 152.

Hawking, S. W., Ellis, G. F. R. 1973, *The Large-Scale Structure of Space-Time* (Cambridge: Cambridge University Press).

Hawking, S. W., Penrose, R. 1970, "The singularities of gravitational collapse and cosmology," *Proc. R. Soc. London*, **A314**, 529–548.

Israel, W. 1967, "Event horizons in static vacuum space-times," *Phys. Rev.*, **164**, 1776.

Kamran, N., McLenaghan, R. G. 1984, "Separation of variables and symmetry operators for the neutrino and Dirac equations in space-times admitting a two-parameter abelian orthogonally transitive isometry group and a pair of shearfree geodesic null congruences," *J. Math. Phys.*, **25**, 1019.

Kerr, R. P. 1963, "Gravitational field of a spinning mass as an example of algebraically special metrics," *Phys. Rev. Lett.*, **11**, 237.

Khan, K. A., Penrose, R. 1971, *Nature*, **229**, 185.

Landau, L. 1932, "On the theory of stars," *Phys. Z. Sowjetunion I*, 285.

Lemaître, G. 1933, "L'universe en expansion," *Ann. Soc. Sci. Bruxelles I*, **A53**, 51. (cf. p. 82).

Lifshitz, E. M., Khalatnikov, I. M. 1963, "Investigations in relativistic cosmology," *Advances in Physics*, **12**, 185.

Mason, L. J., Woodhouse, N. M. J. 1996, *Integrability, Self-Duality, and Twistor Theory* (Oxford: Oxford University Press).

McNamara, J. M. 1978a, *Proc. R. Soc. London*, **A358**, 499.

McNamara, J. M. 1978b, *Proc. R. Soc. London*, **A364**, 121.

Newman, E. T., Couch, E., Chinnapared, K., Exton, A., Prakash, A., Torrence, R. 1965, "Metric of a rotating charged mass," *J. Math. Phys.*, **6**, 918–919.

Newman, E. T., Penrose, R. 1962, "An approach to gravitational radiation by a method of spin coefficients," *J. Math. Phys.*, **3**, 896 (errata **4**, 1963, 998).

Nutku, Y., Halil, M. 1977, *Phys. Rev. Lett.*, **39**, 1379.

Oppenheimer, J. R ., Snyder, H. 1939, "On continued gravitational contraction," *Phys. Rev.*, **56**, 455–459.

Oppenheimer, J. R., Volkoff, H. 1939, "On massive neutron cores," *Phys. Rev.*, **55**, 374.

Page, D. N. 1976, "Dirac equation around a charged, rotating black hole," *Phys. Rev.*, **D14**, 1509.

Penrose, R. 1965, "Gravitational collapse and space-time singularities," *Phys. Rev. Lett.*, **14**, 57–59.

Penrose, R. 1969, "Gravitational collapse: the role of general relativity," *Rivista del Nuovo Cimento, Numero speciale*, **1**, 252–276.

Penrose, R. 1973, *Ann. N.Y. Acad. Sci.*, **224**, 125.

Penrose, R. 1978, "Singularities of space-time," in *Theoretical Principles in Astrophysics and Relativity*, ed. N. R. Lebowitz, W. H. Reid, and P. O. Vandervoort (Chicago: University of Chicago Press).

Penrose, R. 1979, "Singularities and time-asymmetry," in *General Relativity*, ed. S. W. Hawking & W. Israel (Cambridge: Cambridge University Press).

Robinson, D. C. 1975, *Phys. Rev. Lett.*, **34**, 905.

Regge, T., Wheeler, J. A. W., 1957, "Stability of a Schwarzschild singularity," *Phys. Rev.*, **108**, 1063.

Simpson, M., Penrose, R. 1973, *Int. J. Theor. Phys.*, **7**, 183.

Synge, J. L. 1950, "The gravitational field of a particle," *Proc. Irish Acad.*, **A53**, 83.

Teukolsky, S. A. 1973, "Perturbations of a rotating black hole: Fundamental equations for gravitational, electromagnetic, and neutrino-field perturbations," *Astrophys. J.*, **185**, 635.

Walker, M., Penrose, R. 1970, "On quadratic first integrals of the geodesic equations for type {22} spacetimes," *Comm. Math. Phys.*, **18**, 265.

Woodhouse, N. M. J., Mason, L. J. 1988, "The Geroch group and non-Hausdorff twistor spaces," *Nonlinearity*, **1**, 73–114.

Zerilli, F. J. 1970, *Phys. Rev.*, **D2**, 2141.

11

Chandra and His Students at Yerkes Observatory

Donald E. Osterbrock

11.1 Introduction

Subrahmanyan Chandrasekhar spent more than a quarter of a century at Yerkes Observatory, a large part of his scientific career. While he was in residence there he wrote four books and more than two hundred papers, moved up the academic hierarchy from research associate to distinguished service professor, and became an American citizen. Other chapters in this volume summarize and evaluate Chandra's research in the many different fields of astrophysics in which he successively worked; each is written by a distinguished expert in that field. My own chapter is different; in it I try to describe his scientific activities at Yerkes, particularly in teaching, advising, and molding graduate students, of whom I was fortunate to be one. This contribution is therefore based on the memories of a participant, but with very great help from many fellow "Chandra-Ph.D.s" who responded to my requests for specifics of their careers, and of their insights into our former mentor's role in preparing them for independent scientific work. I have tried to follow the goal enunciated by my fellow author, Norman Lebovitz, to analyze Chandra's contributions seriously and as fully as I can, avoiding extravagant praise, of which he was wary, but taking him quite seriously, as he surely would have wished (and as I always did!).

Chandra's biography, by Kameshwar C. Wali (1991), is an excellent record of the events of his life, as he saw it himself in his late sixties and his seventies. The treatment of his Yerkes years as given there is rather brief, however, and I hope that this chapter, written from my own quite different perspective, will add new insights into his very great contributions toward preparing the next generation of research astrophysicists. Although my main focus is on Chandra's years at Yerkes, I continue with his teaching and Ph.D. students on campus in Chicago, after his move there in 1964, until his last Ph.D., Bonnie D. Miller, was awarded her degree in 1973.

11.2 Early history

As described fully in my book, *Yerkes Observatory 1892–1950: The Birth, Near Death and Resurrection of a Scientific Research Institution*, in 1936–37 its young director, Otto Struve, recommended the appointments of the even younger Gerard P. Kuiper, Bengt Strömgren, and Chandra to the University of Chicago's young president, Robert M. Hutchins. All three were foreigners; such appointments were unusual in those days when most scientists in American universities came from families which had been in this country for generations. But Struve wanted the best, wherever he could find them, and Hutchins backed him fully. Struve wanted to make the University of Chicago the outstanding power in astrophysics in the world; Kuiper was an observer whose interests were in that direction, while Strömgren and Chandra were theoretical astrophysicists, a very rare breed in the United States of those years. Nearly all the astronomy department faculty members lived and worked at Yerkes Observatory, in the little village of Williams Bay, Wisconsin, some eighty miles from the campus, a site selected to be out of the smoke, haze and fog of Chicago, and, as it turned out, free of the growing light pollution as well. William W. Morgan, an observational spectroscopist who had been Struve's second Ph.D. thesis student, was already on its staff. He, with Chandra, Kuiper, and Strömgren, became the key members of Struve's brilliant young research group.

There had always been a small outpost of astronomers on the campus in Chicago, devoted entirely to celestial mechanics, and closely connected with the mathematics department. Its most famous member had been Forest R. Moulton in the early years; the retirement of William D. MacMillan in the summer of 1936 created one of the openings for a new faculty member which Struve filled. He wanted to replace the celestial mechanics experts on campus with astrophysicists. Strömgren, when he arrived in September 1936, was originally stationed in Chicago, but he was so productive and valuable in research that after only two quarters Struve moved him to Yerkes, so that he could interact fully with all the other staff members who were working there.

When Struve recruited Chandra, his initial thought had been to put him on the campus with Strömgren. Hutchins was strongly in favor of this idea. But Henry G. Gale, the laboratory spectroscopist who was dean of physical sciences, had been born and grown up in Aurora, Illinois, and spent his entire adult life at the University of Chicago. He was strongly prejudiced against anyone with a dark skin. To him Chandra was a black, Negro, or "colored man," the polite term of his time. Chicago was a *de facto* segregated city, Hyde Park was an all-white suburb close to the boundary of the Black Belt, and Gale would not allow a "Negro" to teach in his division on the campus. Struve, Kuiper and Strömgren, born and raised abroad, were completely free of this prejudice; Hutchins, the son of a liberal Presbyterian minister who had been a professor at Oberlin College, Ohio before he became president of Berea College, Kentucky, had been brought up to abhor it. However, Gale was not alone in his beliefs; the University of Chicago trustees and business agents were determined to keep Hyde Park an all-white enclave to protect its heavy investment in residential property in the campus area, and probably a majority of the American-born faculty members of his division shared his feelings about blacks, although not about a high-caste Indian with a Ph.D.

Struve was aware of all this. He was determined to have Chandra on his faculty, and although Gale advised against the appointment and forbade the Yerkes director to station the Indian astrophysicist on campus, Struve went around him and carried on his negotiations directly with Hutchins and Emery T. Filbey, dean of the faculty. The director carefully paved the way for Chandrasekhar when he came west from Harvard, where he was lecturing, to give two colloquia and see the observatory and its staff. Struve made a reservation for him at International House, the only unsegregated lodging place where a short-term visitor could get a room near the campus, cautioning everyone that Chandra was a distinguished Indian scientist, and a fellow of Trinity College, Cambridge. Struve sent Chandra careful directions on just how to get to the right railroad station in Chicago, had a car from Yerkes meet him at the station near Williams Bay, insisted that he stay as a guest in his home where Struve's wife prepared his vegetarian meals, and drove him back to Chicago himself. President Hutchins did not have time to meet Chandra, but afterward at Struve's suggestion sent him a radiogram on the ship on which he was returning to England, urging him to accept the Chicago offer.

It tipped the balance, and when Chandra returned to Williams Bay to stay in December 1936, Struve arranged for him and his bride, Lalitha, to stay with the Kuipers until their own house was ready for them. The director prepared a general letter of introduction for him, attesting that "Dr. Chandrasekhar of Madras, India and Cambridge University, England" was now "a valued member of our scientific staff," to help smooth the way for him and his wife in the little Wisconsin community.

Chandra was proud and sensitive; needless to say, he knew what was going on, and was well aware of the slights to which he could all too easily be subjected outside of Williams Bay. Even as late as the 1950s, when he and his wife went on summer vacations, he found it expedient to phone ahead to resorts and hotels to explain that they were Indians, to make sure they would be welcome. It was not easy for him to live in America, and it marked his personality (Wali 1991; Osterbrock 1997).

11.3   Courses at Yerkes Observatory

With the coming of his three new faculty members in 1936–37, Struve reorganized the graduate teaching at Yerkes Observatory. He knew from his own experiences as a graduate student there in 1921–23, and as a faculty member since then, that the previous system was woefully inadequate. Typically there had been three to six students, spending most of their time working as assistants and doing research, and taking three reading and research courses each quarter (except in summer, when their number would swell as another five or six teachers at nearby colleges and universities arrived, to work leisurely toward the Ph.D. degrees which would bring them salary raises and promotions at their home institutions). Each professor taught his own specialty, and since there were no theoretical astrophysicists on the faculty, no one taught that subject. Struve, a demon observer who had essentially no training in modern or even advanced physics, knew only what little astrophysics he had been able to pick up by reading, but was determined that the Yerkes students should learn what he had not. He set up a two-year cycle of eighteen one-quarter courses, three to be given each quarter. All the graduate students were required to take all the courses, and their final examination for the Ph.D. was based upon them. Chandra's first teaching assignment was three quarters of stellar interiors, the subject on which he was then working, spread over his first two years at Yerkes. Strömgren taught stellar atmospheres, but in the spring of 1938 he returned to Copenhagen, and from then on Chandra taught that subject too. Struve put him in charge of the graduate teaching program after Strömgren's departure. There were typically ten to fifteen students, most of whom were taking the courses, as the normal time required to earn a Ph.D. was three years after arrival at Williams Bay. By the immediate post–World War II period, when Chandra did the bulk of his teaching at Yerkes, the course structure and schedule were well established. No courses were taught in the morning, so the students and professors who had observed with the telescopes after midnight could try to sleep. Monday afternoon was the colloquium time, while Tuesday, Wednesday, and Thursday afternoons were devoted to one course each. Typically a lecture would continue for an hour to an hour and a half. Friday afternoon any professor who had not finished all he had wanted to say in his regular class period could give a second lecture, or if a long-term visitor such as Pol Swings was presenting a series of lectures, it would be his day. Most faculty members assigned some homework in their courses, frequently based on observational data or observing assignments, and nearly all gave either a long take-home problem or a final examination at the end of the quarter.

All the courses were given in the one lecture room, at the end of the long hall of the observatory building, nearest the dome of the 12-inch refractor. All of the students had bachelors' degrees when they came to Yerkes, many had masters' degrees from other universities, and none of them ever had to go to the campus except to graduate.

11.4 Chandra's teaching

Chandra eventually taught not only stellar interiors and atmospheres, but also stellar dynamics, and at times even molecular spectroscopy (after Gerhard Herzberg, whose course it had been, and John G. Phillips, his student and successor on the faculty, had both left Yerkes). Chandra's lectures in all his courses were formal and highly mathematical, with very little discussion of the physical ideas. He wore a dark suit (or a light gray one in summer), white shirt, and conservative tie, and spoke in complete sentences. Chandra presented the basic equations, such as the equation of radiative equilibrium, and then worked through the methods of solving them, doing all the mathematical steps at the blackboard, only very rarely making a mistake in sign or arithmetic as he transposed terms or factored equations, and seldom consulting his notes. His presentation was very well organized and logical, with no loose ends. Chandra's Indian-accented English was hard for many American students to understand easily at first, and as he proceeded through the equations they often had difficulty keeping up with him. However, they soon overcame whatever language problem they had, and came out of his courses with "good sets of notes," as several of them told me, which some of them used as reference sources for their own teaching for years.

All of the graduate students who did their Ph.D. theses with Chandra were good in mathematics; he selected them for that skill or ability, and would not accept students who were not. Hence it is not surprising that nearly all of them considered him an excellent teacher, from whom they learned a lot. Some wished that he had taught more about the basic physical ideas than about the mathematical steps he had developed to go from them to the final results, but that was his style of doing science, and it took him far indeed.

In my own experience, Chandra was a very good teacher, and at his best in teaching a course which he had given two or three times previously, such as stellar atmospheres and radiative transfer in my time at Yerkes (1949–52). Then he knew the material well, had it organized in excellent fashion, and was still fresh and interested in it. All the courses he taught were in subjects in which he had worked, produced a long series of papers, taken it as far as he could go, and eventually written a book, although that generally came after he had taught it several times. A subject he had taught many times, and had not worked in for several years, such as stellar interiors in my time, was no longer of central interest to him, and in fact perhaps somewhat boring. Then he could not communicate the same interest and intensity, and as he hated to waste time in preparation for a lecture on a subject he had taught so many times, might make a minor slip in an equation and get bogged down in correcting it while he was at the board, rather than dealing with the central issues of the topic. Once Chandra had written his book, he moved on to a new subject and made little attempt to keep up with the developing literature of the old one. In his courses he taught the subject as he had found it, with what he and his students had added to it, but hardly anything about newer results that others had obtained. Thus we learned little of the early groping steps toward understanding stellar evolution which George Gamow had recently made, and which Martin Schwarzschild, Fred Hoyle, and their collaborators were making then. Chandra's way was excellent for preparing students to work with him on the research he was then doing, but was not so good for producing well-rounded research scientists from the students who were working with other professors on more observational topics, but needed to understand the basic ideas of stellar atmospheres and stellar interiors.

However, the majority of the observationally oriented Yerkes students of the 1940s and 1950s who answered my recent queries were quite positive about Chandra's teaching. He did his best to get the basics across, and at the very least succeeded in communicating his enthusiasm to them. Those who were mathematically inclined particularly enjoyed his approach. In his eagerness to teach, Chandra could be demanding, dictatorial, sometimes even insulting with students he did not know well, although in his own mind he was simply trying to impress them with the necessity of more study to achieve understanding. Thus he could reply curtly to questions in class, catechize students he encountered in the hall or library, or otherwise humiliate them in the presence of others. A few of these students were afraid of him; one was known to flee from his basement cubicle when he heard Chandra's characteristic footsteps coming down the stairs from his office on the first floor. Others, in self-defense, replied semi-contemptuously to Chandra (or so it seemed to him), marking themselves as enemies in his eyes. He never subjected his own students to this treatment, nor the great majority of the observational students, but some of those whom he did harass in this way have never been able to forget it. Chandra drove more than one student out of Yerkes Observatory, but surely not everyone who came there should expect to get a degree automatically, he would have replied.

11.5 On the campus

Chandra did not teach a course on the campus until after World War II, although he and several other professors from Yerkes had commuted to the city to give one or two lectures each in an elementary course at the University of Chicago Downtown Center in the spring of 1938. Even this had aroused Dean Gale's ire, but Struve, with Hutchins's support, had faced him down, as described in detail, with full references to contemporary documents, in my book (Osterbrock 1997). However, Gale retired in 1940, to be succeeded by Arthur H. Compton, a socially responsible physicist who had many ties with India. Chandra was never again unwelcome on the campus, and by 1949–50 he was an internationally renowned scientist, whom Chancellor Hutchins (as he was now titled) would have liked very much to have there. Struve had proposed various reorganization plans, beginning in 1941, some of which involved Chandra's transfer to Chicago, but in the end, for one reason or another, he did not make the move until years later. However, in 1948–49 the astronomy department offered, for the first time, a one-year sequence of beginning graduate courses in astrophysics on campus, designed to interest or hold the attention of students who would then go on to Yerkes to complete their training. Chandra, with Guido Münch collaborating, taught the first course in this series, Astronomy 301, Topics in the Theory of Stellar Atmospheres, in the fall quarter. It was to become the source of one of the great legends of Chandra, which President John T. Wilson loved to tell, and which the subject himself apparently enjoyed (Wali 1991). According to the president's version, Chandra used to drive hundreds of miles between Yerkes Observatory and the campus, week after week, to teach a class consisting of only two students, but in the end all his travel and effort was justified, because in 1957 those two students, Chen Ning Yang and Tsung-Dao Lee, jointly received the Nobel Prize in Physics. Later Wilson could have added: as Chandra, their teacher, did himself in 1983.

In fact, however, there were quite a few more students in the class, and neither Yang, who was then a postdoc working with Enrico Fermi, nor Lee, a graduate student doing the same, was actually registered for the course. They were both sitting in on it, as were Fermi, Marcel Schein (another physics professor who specialized in cosmic-ray research), and several younger physics faculty members, postdocs, and graduate students. Chandra was already a famous theorist and somewhat of a figure of mystery on the campus; a few of the auditors no doubt simply wanted to see what he was like. Their numbers decreased as the quarter wore on, but Fermi, Schein (who often fell asleep in the front row and snored audibly, to Chandra's clear but never vocally expressed distaste), Yang, and Lee remained true to the end of the quarter, probably along with several others whom I can no longer clearly remember. There were actually six students registered for the course to the end, as the grade record in the files of the Office of the Registrar, signed by Chandra, shows. Three of them, Richard L. Garwin, who was then also doing his thesis with Fermi, Arthur Uhlir Jr., later a professor at Tufts, and I, went on to Ph.D. degrees at Chicago; a fourth, John Goddard, died not long thereafter, before finishing his degree. Garwin, Uhlir, and I certainly attended the classes faithfully, and learned a lot from them, and I think that Goddard did too. Wilson's story clearly shows the fallibility of human memory long after an event, and the value of contemporary records in establishing facts. Yang's and Lee's recollections also agree with mine (and with the records); Yang told Walter Sullivan, the writer of Chandra's obituary for the *New York Times*, that there had been such a class but there were more than two people in it. The other two students who had registered for the course in addition to the four of us apparently did not complete graduate degrees at Chicago, and may have stopped attending Chandra's lectures; in 1995 the Registrar, bound to respect the privacy of students' records, could only write me that some of the grades "were not satisfactory."

However, Lalitha told me in 1997 that Wilson had embroidered the story, and that his version was not correct. G. Srinivasan, the editor of this volume, confirmed that Chandra had related a different version to him in 1967, which in its main lines was the same as the one Lalitha told me. No doubt it is what really happened. *One* day during that winter quarter of 1948–49, there was a severe blizzard which closed the roads in Chicago and southern Wisconsin. Nevertheless, Chandra fought his way to his class, mushing through the snow to the commuter train, which reached Chicago, then made another hike through deep snow from the Northwestern station across the Loop to the Illinois Central station at Randolph Street and out to the campus, and made a third foray on foot to Eckhart Hall. He arrived there just five minutes before the class was scheduled to start, and only Yang and Lee were present. He gave the lecture to them, and managed to get back to Williams Bay that evening, following all the steps in reverse. According to Lalitha, that morning Chandra's colleague Kuiper had urged him not to try to make his way to the campus, but he had persisted. When Chandra returned to Yerkes Observatory that night and revealed that only two "students" had showed up for his class, Kuiper joked sarcastically at his expense for wasting a day and risking being marooned somewhere along the way, to give a lecture for which only two auditors were present. Thus when Yang and Lee received their Nobel Prize eight years later, Chandra took great pleasure in reminding (or telling) Kuiper that they were the two for whom he had made that exhausting trip to Chicago, and it *had* been worth it after all.

Very probably it happened that way; no doubt the University of Chicago officially cancelled all classes that day, and only Lee and Yang came to the classroom because they had learned, possibly from Fermi, that Chandra would nevertheless be on the campus. Wilson's more colorful version is the one which was widely repeated, however, especially in introducing Chandra on public occasions, and it is the story Wali published in his book.

As time went on, Chandra's research moved from stellar interiors, through galactic dynamics, then stellar atmospheres, to turbulence, hydrodynamic and hydromagnetic stability and relativistic astrophysics, and the mathematical theory of black holes and colliding plane waves. After

Struve's resignation and departure for Berkeley in 1950, Strömgren returned as director. He was one of the leading theoretical astrophysicists in the world, working on stellar atmospheres, stellar interiors, and stellar evolution, and on the observational application of these subjects to gaining physical understanding of how real stars form, live, transform themselves into red giants, planetary nebulae, white dwarfs, or supernovae, and die. Strömgren was up to date in all these fields and it was natural for him to take over teaching them. Chandra was no doubt glad to let him do so, to free his time for his own research, in which he tended more and more to emphasize mathematical beauty and elegance. Once he gently chided me, probably during the 1963–64 academic year when I was a visiting professor at Yerkes, for spending so much time on planetary nebulae and H II regions. Everything I was doing depended on observational data, he told me, which could easily turn out to be wrong. (I did not think it wise to mention that much of that data actually came from my own observational work, and the work of my Ph.D students of the time!) His own work, Chandra said, was based on a few easily stated assumptions, and, being mathematical, would always endure. To most of the graduate students who then were coming to Yerkes Observatory, learning about the real universe seemed more attractive than mathematical truth and beauty. Many of the faculty members shared that feeling. Under these circumstances a growing sense of alienation naturally arose between Chandra and his colleagues, as is well expressed in Wali's book.

With the diplomatic Strömgren at the helm, Chandra could continue to work effectively at Yerkes, but his interests were turning increasingly to the campus in Chicago. His two papers with Fermi on magnetic fields in the interstellar matter in the spiral arms of the Galaxy and their stability played a powerful role in this attraction (Chandrasekhar and Fermi 1953a,b). Chandra deeply admired this outstanding genius of physics, whose philosophy was "to use every dirty trick at your command" (combining theoretical and experimental reasoning) to solve the most important physical (and hence astrophysical) problems of his time. By then Chandra's own approach was quite different. He had tremendous mathematical powers. Guido Münch, who worked closely with Chandra for several years, commented on one aspect of this. He could work for weeks on the solution of a complicated equation or set of coupled equations, and in the end break it, often guided by his "intuition," actually the result of years of concentrated experience.

11.6   Observational research

Full-time theoretical astrophysicists were exceedingly rare in America when Struve hired Chandra on the Yerkes staff in 1936. Henry Norris Russell at Princeton was practically the only one who did no observing himself; his Ph.D. thesis student, Donald H. Menzel, who had joined the Harvard faculty in 1932, was the other, much younger, theorist in the country, and he did observational work as well. Struve was convinced that America in general and Yerkes Observatory in particular could only really advance with first-class theoreticians on the scene, to guide the observers' thinking and interpret their results, and he went abroad to get them. Many working astronomers were highly skeptical, and Harlow Shapley, director at Harvard, had advised Struve against adding Chandra to the Yerkes faculty (although he was simultaneously trying to persuade the young theoretician to accept a further short-term appointment at his own university). Even Struve had some doubts about a "pure" theoretical astrophysicist, and wrote Kuiper, who was then at Harvard with Chandra, that it would be "decidedly advantageous" if the latter would undertake "a small amount" of observational work at Yerkes. When Kuiper mentioned this idea to Chandra, he welcomed the idea in principle, and thought he might try some observational work on solar prominences, connected with his theoretical ideas on the outer layers of the sun.

Undoubtedly Chandra was sincere in this thought, but research is a highly specialized business, and he never had the time to learn all the intricacies of operating a large refracting telescope and a solar spectrograph. It would have been a great waste of his talents, as Struve recognized as soon as Chandra got to Yerkes and began producing papers and books packed with new theoretical results. The director never brought up that idea again.

But Chandra coauthored one purely observational paper, years later, surely of his own volition. This was a report on an eclipse expedition, on which he photographed the solar corona rather than the chromosphere or prominences. The eclipse had a relatively short totality, only thirty-seven seconds, but it occurred in July 1945, just as World War II was winding down to a close, two months after the defeat of Germany and two months before the Japanese surrender. All during the war Chandra had been working hard on weapons development at the Army Ballistic Research Laboratory at Aberdeen, Maryland, alternating three weeks there and three weeks back in Williams Bay, where W. Albert Hiltner, the leader of the eclipse group, had been working on the same type of project at the Yerkes Optical Bureau. No doubt the trip to the remote observing site on the line of totality near Pine River, Manitoba was a welcome diversion after three years of wartime tension. They were joined there by Burke Smith, a stellar spectroscopist who had collaborated with Struve at Yerkes. Chandra and Smith helped Hiltner set up the two photographic telescopes they had brought with them from Yerkes. On the eclipse date the sky was clear and Chandra got a good photograph of the corona with the shorter-focal-length instrument, while Hiltner obtained two of the outer chromosphere and inner corona with the large-scale instrument. They duly published reproductions of these photographs in a two-page paper, in the tradition of the time (Hiltner and Chandrasekhar 1945), and Chandra soon went back to his important theoretical research on radiative transfer.

A second paper Chandra wrote described briefly some of the earlier observational work at Yerkes. It was one of a series of articles written at Struve's behest, each by one of the top research workers at Yerkes, for its semicentennial in 1947, to be published as a group in *Science* fifty years after the observatory's dedication. Each article described the earliest work by Hale, Walter S. Adams, and the other giants of the past in one particular subject, and emphasized the continuity by which that work had led down to the writer's current research. This was easy enough for Struve, Morgan, Kuiper, and Gerhard Herzberg, the molecular spectroscopist who had recently joined the Yerkes faculty, but there had been no theoretical astrophysicist at all in 1897 with whom Chandra could connect his work. In 1947 he was in the midst of his radiative-transfer and H^- period, so he wrote his article on "Solar research and theoretical astrophysics" (Chandrasekhar 1947). It began with Hale's spectroscopic confirmation, made at his own Kenwood Observatory in Chicago, that the yellow D3 emission line in the chromosphere was truly the same line emitted by helium gas on the earth, discovered in 1895; his later discovery of C_2 in emission in the low chromosphere; Edison Pettit's studies of solar prominences; and Philip C. Keenan's observational work on the solar granulation, all done with the 40-inch refractor. Then Chandra smoothly switched to his and his students' theoretical work on H^-, the continuous spectrum of the sun and its limb darkening, and scattering by free electrons in the atmospheres of hot stars. They all "serve[d] to underline a fact which Hale often emphasized," Chandra concluded, "namely, that there is *no essential difference between the attitudes of a physicist and an astronomer.*"

Years later, in 1983, Case Western Reserve University awarded Chandra its Michelson-Morley Prize. Peter Pesch, a former Yerkes observational Ph.D. and a faculty member at CWRU, introduced Chandra for his prize lecture there. In his introduction Pesch showed a slide of the first page of the Hiltner and Chandrasekhar eclipse paper, listing the great relativity theorist as one of its coauthors, and another slide reproducing the coronal photograph he had taken. Chandra, picking up on the joke instantly, started his lecture with the comment that Pesch had "destroyed [his] credibility!"

11.7 Colloquia

Struve put Chandra in charge of the colloquia at Yerkes, and he remained in that post until very nearly the year when he moved to Chicago. Monday afternoon was colloquium time, and one was scheduled every week as regularly as clockwork. Naturally there had been colloquia before he joined the Yerkes faculty, though on a more catch-as-catch-can schedule; he ignored them and began numbering the colloquia from the day his reign began, like an ancient king or emperor. There were plenty of scientific visitors to Williams Bay in Struve's and Strömgren's years as director, and Chandra saw that they all gave colloquia on their current or recent research while they were there. Gaps in the visitors were filled in with specific invitations to astronomers from nearby Madison, Northwestern University, the University of Illinois, and other research centers, and to physicists from the campus. The senior Yerkes faculty members generally each gave one colloquium a year; younger assistant professors and instructors, and sometimes even graduate students, also gave them occasionally.

Chandra followed the English tradition of holding a regular tea after the colloquium, and in those bad old sexist days of my youth the faculty wives acted as hostesses, one scheduled for each week, pouring the tea and providing the refreshments. They vied with one another in baking and bringing rich cakes and cookies, which we students wolfed down whenever we thought Chandra's back was turned, though he seemed to enjoy them too. The colloquia were held in the classroom down the hall from the library, and he had his own special chair, and his own special place, at the end of the second row nearest the door. Occasionally an unwary visitor would sit there before Chandra appeared to claim his seat. When he did come in a moment later, usually with the speaker in tow, he would recognize the situation at a glance and take another chair, still empty in the first row, and place it halfway outside the door, next to the visitor, and sit down there. The visitor, now embarrassed, would try to give up his chair, but Chandra would not hear of it; his politeness would call forth further apologies and protestations from the visitor. Finally, after three or four offers and refusals, Chandra would at last accept his chair and the talk could begin. After the tea, one of the students was assigned to wash the colloquium china cups, saucers and spoons, usually done in one of the basins in the men's or women's room, or occasionally in the one bathtub in the building.

We all certainly learned a lot of current astrophysics in these colloquia, perhaps imperfectly, but at least the central ideas. And Chandra made sure we were there to learn it; every graduate student and every faculty member was expected to attend every colloquium, and any student with the temerity to skip one, even it if was only "Recent Research at Such-and-such Observatory" by a visiting director who was more of an organizer than a research scientist, was sure to be subjected to a searching cross-examination the next day.

Chandra himself gave at least one colloquium every year, and frequently more. He always gave the even-hundred-numbered colloquia, making them festive occasions on which Lalitha, his wife, poured the tea and provided a special cake, but he always had a serious scientific message to bring to the auditors. His colloquia, like his lectures, were models of organization, extremely well presented and always interesting. In his early, pre-war years,

Chandra gave even more colloquia, many of them didactic, for the faculty members as well as the students. Theoretical astrophysics was a subject few of them had studied, and he widened their horizons.

11.8 Seminars

In addition to the colloquia, Chandra ran a theoretical seminar during much of his period at Yerkes. He began it a few years after World War II, when the great dammed-up wave of new and returning graduate students hit Yerkes Observatory and provided him with a steady source of good students. Most probably his seminars began operation in the summer or fall of 1948, when Chandra was seriously examining published work in turbulence, a subject astronomers, led by Struve, believed they had discovered empirically in stellar atmospheres years before. After the war such leading theoretical physicists and aerodynamicists as Werner Heisenberg, C. F. von Weizsäcker, and young G. K. Batchelor had begun publishing papers on it, whetting Chandra's interest. In later years he went on to magnetohydrodynamics, then to rotating ellipsoids, and then to general relativity.

The seminar was held regularly in the classroom on Monday evenings, and Chandra expected all the theoretically oriented graduate students who wanted to work with him (the two groups were identical at first, as there were no other senior theorists until Strömgren's return in 1951) to be there, as well as any postdocs or visitors who were theorists. These seminars were Chandra's way of getting into a new subject, and keeping abreast of the latest work in the field he was working on. He would assign papers, some recently published, and others which he had received as carbon copies of manuscripts just submitted for publication, in those pre-Xerox, prepreprint days. Often he would report on the most interesting new papers himself. Whoever was assigned the paper was expected to study it in depth, work through all the equations, look for good new ideas and also for weak points, and report orally on it in two or three weeks. Ideally the report was very thorough; those in attendance were encouraged to ask questions, as Chandra himself always did.

He gave many of the reports himself, at least in the years I was there, when he was already working on turbulence and was starting to get into magnetohydrodynamic and plasma problems. In this situation he was frank in mentioning problems he had in understanding what an author was trying to do, and would welcome comments, questions, and suggestions. He was happiest when he uncovered an error, found a mathematical shortcut the author of the paper had not seen, or in the course of analyzing and discussing the paper formulated a new problem which would be grist for his mill, or for his students'. More than once his report on a paper gradually changed, over a period of a few weeks, into outlining a new paper he was writing, going beyond it or straightening out some of the flaws in it.

These seminars were an excellent introduction to actual research, for students who had previously been totally immersed in undergraduate or beginning graduate course work. Chandra was demonstrating how a real theorist works, welcoming our comments and questions, never answering curtly or abruptly, as he sometimes did in class lectures, when he was frequently under time pressure to keep up to his planned pace. T. D. Lee spent two quarters at Yerkes, the spring and summer of 1950, after completing his Ph.D. thesis on white-dwarf stars under Fermi on the campus, with Chandra as the astronomical consultant. The brilliant young postdoc (then twenty-three) attended the theoretical seminars regularly, and played a prominent part in the discussions. Their styles contrasted greatly, Chandra much more mathematical in his approach, Lee more physical, and when they occasionally reached different conclusions, groping toward an understanding of turbulence, the fur could fly. But they both remained civil, and the next day would again be discussing whether the mean turbulent kinetic-energy density was approximately equal to the mean turbulent magnetic-energy density (Lee's formulation) or to a mean-square expression involving the curl of the magnetic field (Chandra's result). At times it seemed to be almost a replay of the arguments between Eddington and the brilliant young Chandra over the internal structure of white-dwarf stars a decade and a half earlier, now reenacted in the quiet halls of Yerkes Observatory instead of at the Royal Astronomical Society's meetings.

Chandra's Henry Norris Russell Lecture (the third ever given, following the first one by Russell himself, and the second by the recently retired, great dean of observational astrophysics, Walter S. Adams) was on turbulence, nearly all of it based on material he and his students had discussed in that first year or two of the seminars (Chandrasekhar 1949). His graduate students were stimulated by the seminar series as well; two of them, after receiving their Ph.D.s, went on studying and developing more applied aspects of turbulence theory, Marshal H. Wrubel as a postdoc at Princeton, and Su-shu Huang as a research associate who stayed briefly at Yerkes before following Struve to Berkeley (Wrubel 1950b; Huang 1950). My own little theoretical paper on the contribution of elastic scattering of free electrons by neutral H atoms to reducing the electrical conductivity of the solar atmosphere came out of the beginning magnetohydrodynamics period of the seminar series (Osterbrock 1952b).

11.9   Computers

Chandra's research depended on large amounts of numerical computing, especially numerical integration of differential equations. He, like other theorists of his time, was an expert in carrying out such calculations, using an electric-powered, hand-operated computing machine. His graduate students learned to do it too, and several of them worked as assistants for him, especially in their earlier years, doing the time-consuming numerical work. In this research, separate from their thesis problems, they might participate in the theoretical development to a certain extent, but spent most of their efforts on computing. Often they became coauthors of the resulting papers. An example is some of the early work Chandra did on H$^-$ with Margaret Kiess Krogdahl (Chandrasekhar and Krogdahl 1943). It helped support her, and at the same time prepared her for her thesis with him on the inhomogeneous Stark effect in stellar atmospheres (Krogdahl 1944a,b).

However, for maximum long-term efficiency in Chandra's ongoing research, a full-time computer (the name then used for the person who used the machine) was clearly preferable. Theodosia ("Theo") Belland, a resident of the nearby village of Fontana, became his first full-time computer, from 1940 to 1943. Earlier she had worked for Struve, and her husband, Fred Belland, was also on the observatory staff. Chandra himself taught Theo Belland how to carry out all the steps necessary to integrate numerically whatever complicated definite integrals, differential equation, or system of equations he had derived, as he later did for his other computers. In 1944 Frances Herman succeeded to the post; soon afterward she married and became Frances Herman Breen. She worked with Chandra through 1948, but resigned when she was about to have her first child. In early 1949 Donna D. Elbert, like Frances Breen a graduate of Williams Bay High School, took the job. Chandra included both of them as coauthors on papers for which they did unusually large amounts of numerical work. Donna Elbert was to continue with Chandra for more than three decades, and to become an outstanding numerical computer.

11.10 The Astrophysical Journal

A prolific author, Chandra published most of the papers he wrote at Yerkes in the *Astrophysical Journal*. Ultimately he published 137 papers in it, up to 1994 a record second only to that of Struve, who published 228 papers in it in his lifetime (Abt 1995b). Like many enduring astronomical institutions, the *Astrophysical Journal* was the result of Hale's organizational activities, founded by him and his older, then better-known astrophysicist friend, James E. Keeler, in 1895. The *Journal* belonged to the University of Chicago, and was published by its Press; Hale and the successive directors of Yerkes Observatory after him, Edwin B. Frost and Struve, were automatically its managing editors. From its start the *Astrophysical Journal* was the leading journal of astrophysics in America, with nearly all the papers from Yerkes and, after its founding, Mount Wilson Observatories, published in it, and the astrophysical papers from most other observatories and research centers in the United States. There were many from abroad. After 1942 the *Journal* was published "in collaboration with the American

Astronomical Society," and Harvard and Lick Observatories, previously the main holdouts, also sent their papers to it, but the University of Chicago Press retained ownership, control, and the managing editorship. After World War II ended, Struve was tired and overworked; in 1946 he named Chandra associate managing editor, and then in 1947 gave up the managing editorship to Morgan. No doubt Struve still believed that an observer should hold that post, rather than a theoretician. After Struve's departure in 1950, Chandra played the leading role in negotiating an agreement with the AAS, under which it gained more control over the editorial board and policy of the *Journal*, in exchange for the financial assurance the Society provided by requiring its members to subscribe to it. Chandra remained associate managing editor under Morgan, but the latter suffered a nervous breakdown, was hospitalized, and resigned the editorship in 1952. Then there was no choice but for Chandra to replace him, and he continued in the post for nineteen very fruitful years.

His first editorial assistant at Yerkes, where the papers were received, acknowledged, and sent out to referees, and where he accepted or rejected them (or, more frequently, returned them to their authors for revisions), was Mary Horvath Richmond. She, like Frances Breen and Donna Elbert, was a locally recruited Williams Bay woman.

With all the copyediting, illustrations, make-up and other technical aspects of the *Astrophysical Journal* concentrated at the Press office on the campus, Chandra had another reason for going to Chicago frequently. When he had made only occasional trips there, he often rode the train, a commuter line to the Loop, but, like Struve before him, he found driving his own car was much more convenient, especially in giving him the freedom to return to Williams Bay late in the evening. Lalitha often accompanied him, and he was always willing to take students or visiting scientists, up to the capacity of his car. He enjoyed company and conversation on the two-hour drives each way. The riders had to be sure to meet him right on time at the appointed corner on the campus for the return trip; Chandra made it clear that he would follow his schedule, no matter if they were there or not, and no one wanted to test him. Accustomed to rising early, he would leave Williams Bay at 6:30 A.M. or so to have a full day on the campus. This caused problems for many of the observational types (and some theoreticians as well) who tended to stay on a schedule of working until well past midnight in their offices, and not arising until just before lunch. On more than one occasion Chandra was pleasantly surprised to find a bright-eyed passenger like Leonida Rosino (a visiting astronomer from Padua) or Imam I. Ahmad (a graduate student from Egypt) waiting early for him in the dark morning as he drove up to the observatory to meet them for the trip to Chicago, not realizing that they had decided it was not worth going to bed for only a few hours, and had stayed up all night.

Chandra worked very hard on the *Journal*, spending increasing amounts of time and effort on it as it grew under his watchful supervision. Struve, in his fifteen years as managing editor, and Morgan, in his five, had taken a broad view of astrophysics and had welcomed papers reporting observational results in the rapidly expanding "new" wavelength regions: radio-frequency, infrared (with sensitive new solid-state detectors), and ultraviolet (based on captured German rockets, which carried small telescopes and spectrographs above the earth's atmosphere). They had also welcomed new theoretical ideas, Struve more warmly than Morgan, but Chandra's long term as managing editor began as the post-war expansion of research science, fueled by massive new government funding, was just taking hold. He was the ideal person to ride it to success, highly receptive to observational papers which he believed to be good ones, and casting a wider net for theoretical papers than any of his predecessors had. There was really no place else but the *Astrophysical Journal* for the now rapidly expanding generation of trained American astrophysicists to publish their papers, and Chandra's tremendous reputation encouraged many physicists to send their forays into astronomy to him rather than to the *Physical Review*. His great self-confidence and wide circle of contacts within the scientific community enabled him to make quick judgments as to whom to ask to referee a particular, newly submitted manuscript, and which reports to trust. He personally refereed many of the theoretical papers, and on occasion decided on the spot to accept what he thought of as especially important new observational papers. Occasionally he made a mistake (he tended to categorize certain theorists and observers as "good" or "bad," and it was hard for him to change his thinking about them), but only rarely, and not too many people were badly hurt.

The *Astrophysical Journal* grew and flourished under Chandra's management. The late 1950s were the beginning of the post-Sputnik era in American science. The nation was prosperous and apprehensive about the U.S.S.R.; money flowed into space research and into its basic background, astronomy and astrophysics. In 1954 Chandra established the *Astrophysical Journal Supplement Series*, for less expensive publication of papers containing larger amounts of tabular data; it evolved into the preferred medium for longer papers. Then, in 1967, concerned that the *Physical Review Letters* was draining off short papers reporting "spectacular new advances in astronomy" which some physicists persisted in sending to it for quick publication, Chandra founded a new, separate *Astrophysical Journal, Letters to the Editor*. It had its own fast-track schedule and was a spectacular success, recapturing the hot discovery papers to what Chandra considered their rightful place. During his reign as managing editor he greatly increased the rate of publication of research results, moving from one issue every two months to two per month; the total number of pages published grew fivefold in his nineteen years as managing editor. In 1970 he set up

the *Journal's* own production manager's office in Chicago, to take some
of the administrative effort off his own shoulders, and to make the task a
little less onerous for future managing editors. Jeanette R. Burnett was
the first holder of the post, while Jeanne Hopkins was the long-time chief
technical editor and copyeditor of the *Journal*. She compiled the *Glossary
of Astronomy and Astrophysics* (with a foreword by Chandra), which went
through two editions as the copyeditors' and publication secretaries' bible.

Chandra, with his handsome, boyish charm, his well-dressed appearance, his unfailing courtesy to women, his enthusiasm, and his generous
praise for a job well done, was an idol to them and to all the Press technical
employees. He commanded their respect, and they were always ready to go
the extra mile for him to get an issue out, if he asked them to. On the other
hand, to complaining authors or recalcitrant referees he could be caustic;
he terrorized more of them than he charmed. His supreme self-confidence,
presence, scientific reputation and rapier-like wit made it impossible to win
an argument with him. Yet for authors who met his exacting standards
and whose work he respected he provided fast, efficient publication. His
one failing as an editor, I thought, was his prejudice against Ph.D. theses
condensed and rewritten into papers and submitted to the *Astrophysical
Journal*; this was contrary to the Yerkes tradition, in which a thesis was
written from the start as a manuscript for publication. Chandra could be
brutal to a first-time author who, he thought, had not cut his thesis down
enough and was trying to slip a padded manuscript past him. Wali's book
recounts a few such episodes from the editor's point of view, but for a young
Ph.D. they could be traumatic.

By the late 1960s Chandra was tired of the job he had done so long and
so well; he was ready to hand it over to a successor whom he could trust.
He was greatly concerned about who this would be, and offered the post to
at least four well-known research astrophysicists whom he had personally
selected. (I was one of them, as Wali has revealed.) It was a daunting
prospect for anyone, because by then the *Journal* clearly absorbed so much
of Chandra's time, effort, and resilience. How could any mere mortal carry
on after him? But Helmut A. Abt accepted the challenge and proved a
worthy successor, who worked hard and improved the *Journal* still further,
over an even longer term than Chandra's.

Before Chandra had given up the managing editorship, he brought
about the transfer of the *Astrophysical Journal* from the University of
Chicago to ownership by the American Astronomical Society. He was convinced that the leading journal in its field could no longer be the property
of a single institution. It was too big, too expensive, and too important.
Chandra personally negotiated an agreement under which the transfer was
effected in 1971. Only his immense prestige in the University of Chicago
and in the astronomical community made this step possible, and he had to
work very hard to bring it about even so. But he succeeded, as he did in

everything he wanted to do (Abt 1995a,b; Osterbrock 1995).

Long before he resigned as managing editor, Chandra had moved to Chicago. His interests had been steadily shifting from theories which applied to readily observable effects in common types of stars to questions of hydromagnetic stability and of rotating ellipsoids. He felt more and more kinship with the physicists on the campus, and less and less with the astronomers at Yerkes, who did not particularly want him to teach his current specialties to their students. The *Astrophysical Journal* necessitated frequent trips to Chicago. All these reasons combined to make the move inevitable. In the years immediately after World War II, he drove to Chicago nearly every week, usually on Thursday so that he could attend the physics colloquium in the late afternoon. Then when he took over as editor he began staying overnight, usually at International House, and spending Friday on campus as well. He started teaching physics there, at first the regular graduate quantum mechanics or electrodynamics course. At least one of the senior physics professors thought that Chandra did not teach enough quantum mechanical applications, and tried to make up for it when he himself taught the next quarter of the course, but all the physicists were glad to have their astrophysical colleague in his office in the Institute for Nuclear Studies on Ellis Avenue. In 1959 he and Lalitha rented a small apartment near the campus, so that they could stay overnight in their own base there, and come more frequently when it suited their plans.

Although it made all kinds of practical sense to move to Chicago and work full time on the campus, it must have been a long struggle in Chandra's mind whether to give up on astronomy at Yerkes Observatory, just as in 1951–53 it had been a hard decision for him to become a U.S. citizen (Wali 1991). In 1964 he published a curious article, "The case for astronomy," quite unlike his typical research papers. This one, which he presented orally at a meeting of the American Philosophical Society in April 1963, was highly nonquantitative, discussing the relations between physics and astronomy in extreme generality. According to his analysis, studies of physical sciences are carried out at two levels: a primary one, seeking to formulate general laws, and a secondary level, seeking to analyze and interpret particular complexes of phenomena in terms of these basic laws. Then he gave several examples, starting with Newton's law of gravitation (primary) and its application, by Newton himself, to interpret Kepler's laws of planetary motion. This illustration showed, he wrote, that the primary level was "the domain of physics as commonly understood," while the secondary level was "the domain of the various special branches of the physical sciences;... astronomy is one such branch." Nevertheless, he gave several examples of particularly important "secondary" analyses: of white-dwarf stars, solar limb darkening, H^-, stellar energy-production, synchrotron radiation, and nonthermal radio emission, ranging from his own earlier work to some of the most relevant astrophysical applications of 1963. Chandra continued that "the *only* crucial empirical evidence for the aesthetically most satisfying physical theory conceived by the mind of man—Einstein's general theory of relativity—[was] the astronomical one derived from the motion of Mercury." This led him to conclude that "the principal case for astronomy is the same as the case for any of the physical sciences. No less: but, perhaps more; for only in the scales provided by astronomy can we discern the largest in the natural order of things" (Chandrasekhar 1964).

But apparently that "principal case" was not enough to keep him at Yerkes Observatory, for that same year he and Lalitha went all the way, and moved to a high-rise building on South Lake Shore Drive. Their apartment was on the north side of the building, with a clear view all the way to the Loop, and Chandra liked to keep a pair of binoculars next to his chair, so he could read the time from the huge clock on the Wrigley Building. Three years later they moved back to the edge of the campus, in a modern, high-security apartment building on Dorchester Avenue, close to International House. It was a walk of only a few blocks to his new office in the Laboratory for Astrophysics and Space Research, where he moved as soon as it was built. They lived in their Dorchester Avenue apartment until his death.

After Chandra moved to Chicago, four successive directors of Yerkes Observatory, Hiltner, C. Robert O'Dell, Lewis M. Hobbs, and D. A. Harper, kept his office there unoccupied and waiting for him for more than twenty years. No one else was assigned to use it, and only rarely, many years after he was gone, was even a short-term visitor allowed to occupy it. Likewise, for the first three years after he and Lalitha moved to Chicago, their university house, once E. E. Barnard's and after him occupied by Frank E. Ross and his family, was kept vacant for their possible use. Chandra came back to Yerkes for departmental faculty meetings, but as they shifted to Chicago he appeared at Williams Bay less and less frequently, usually on a Sunday and often with a visitor who wanted to see the observatory. Finally in 1989 Chandra's former office was turned over to a staff engineer for a year, and since 1991, James W. Gee Jr., who became the Yerkes manager then, has occupied it.

11.11   Chandra's Ph.D. thesis students

Chandra was an outstanding research scientist, recognized by membership in nearly every elite honorary scientific society to which an astrophysicist might conceivably aspire, and by every prize, medal, and award right up to the Nobel Prize. But in addition he was a great teacher, particularly as a thesis adviser or supervisor for more than thirty years. For his own Ph.D. students, he was an outstanding teacher, mentor, opener of wide new horizons, and supporter. He liked to have bright, mathematically inclined students working with him, and I have yet to find one who remembers being his student as anything less than a wonderful experience and training for the future. He made them all work hard, but, looking back on it, they all thought it had been good for them.

In his younger days Chandra's reputation as a great theorist and teacher was not yet made, and theoretical astrophysics did not enjoy the importance it does today. His first Ph.D., I believe, was Gordon W. Wares, who had been an undergraduate student at the University of Washington, and then a graduate student at the University of California. He had spent three years at Berkeley and completed all his course requirements for the Ph.D., but theoretical astronomy still meant celestial mechanics and orbit determination there. Wares wanted to become a theoretical astrophysicist and transferred to Yerkes. Louis R. Henrich was a student at Columbia with the same aim; his professor, Jan Schilt, told him he should go to Chandra and he did. Wasley S. Krogdahl, his third Ph.D., had an excellent undergraduate record on the campus in Chicago, and came to Yerkes by that route.

In 1945, after World War II ended, the big rush of former students back to graduate schools began, and of new ones whose academic careers had been interrupted. Chandra demanded a high degree of mathematical preparation, but he wanted thesis students, and he was a realist. Henry G. Horak remembers that in one of his classes in that period, Chandra, looking at the students, remarked, "I don't think that you're very good, but you're the best that there are." His comment was "quasi-humorous," but it expressed his feelings well; few students were as well-trained and expert in mathematics as he was, but he took the best he could find and made the most of them. By that time Chandra would only accept a student to work with him whom he had already taught at Yerkes and who had done well in that class, or who had other credentials of mathematical skills and strong work habits. Horak had done a master's thesis at Kansas on the application of vector methods to orbit theory, which gained him Chandra's respect.

When I came to Yerkes in 1949 with a master's degree from the campus, Chandra knew that I had done well in his course there the previous year, and on the combined physics-astronomy "basic examination" of that time. D. Nelson Limber arrived from Ohio State University a few quarters later with glowing recommendations from Geoffrey Keller, his professor there, a theorist whom Chandra knew well. No doubt his other thesis students of that era had similar recommendations or backgrounds which convinced him to take them on. He simply assumed that any mathematically oriented students at Yerkes would want to work with him, and until Strömgren returned as director in 1951 nearly all of them did. A decade later, when Maurice Clement wanted to do his thesis with Chandra, the pool of applicants was larger. Clement had to prove himself first, by computing a Cowling stellar model using an electrical hand-calculating machine just before Yerkes bought its first IBM 1620 electronic computer. He passed with flying colors, as much for the excellent grammar and sentence construction of his written report (always a point of pride for Chandra himself and an absolute requirement for his students' theses) as for the correctness of his numerical calculation, which his new major professor had expected but wanted to see confirmed. One later would-be thesis student found an error by a factor of two near the end of one of Chandra's own papers, and thus proved that he was worthy to work with him.

Esther Conwell was Chandra's first Ph.D. thesis student in physics, rather than astronomy and astrophysics. She started working with him in 1945, on improved wave functions for the negative ions O^- and H^-. She had taken all her course work on the campus, and had not even met Chandra before it was time for her to start her thesis. World War II was still in progress, and he was glad to have a thesis student who would not be drafted. She spent less than six months at Yerkes, driving back and forth with him between there and Chicago occasionally, and then completed her thesis and received her Ph.D. after she had moved on to a teaching post at Brooklyn College, New York.

In the early post-war period many of the male students were veterans, supported by the "GI Bill of Rights," which paid tuition and living expenses. There were not many other sources of support, except a few graduate-student assistantships. Struve was generally willing to allot one to Chandra, for a student to do computing work for him, but reserved the rest for observers with the 40-inch refractor. Thus some of Chandra's women students, like Merle E. Tuberg and Marjorie Hall Harrison, had to work several nights a week with the big telescope while they were doing their Ph.D. theses. It was no easy task to "reverse" the telescope, pushing it around the pier from one side to the other, particularly on cold winter nights when the oil was stiff and the observers' heavy insulated flying suits made moving awkward.

After about 1954 most of Chandra's Ph.D. thesis students were in the physics department. Generally when they started to work with him they would stay on the campus but commute to Yerkes on Monday, driving together in a car. He was almost certain to be there for the colloquium, and in the evening they would take part in the seminar before driving back to Chicago late in the evening. But when they began working seriously on their theses, he preferred that they move to Williams Bay so he could discuss their work more frequently than he could on his busiest day there, or on his only day on the campus, Thursday. The physics students who moved to Yerkes were even more bored in the little Wisconsin village than the astronomy students, for they did not get much out of the colloquia, nor sit in on the specialized courses on stellar spectroscopy, radiative transfer, galactic structure, and the like. One of Chandra's physics students from that time vividly remembers his arrival in Williams Bay, when he asked one of the long-term Yerkes students what you did there all winter. He got the one-word answer, "fidget!" Another, Fred Bisshopp, was commuting for one or two days a week for several months, because he was having difficulty finding a landlord who would let him keep his dog, a boxer named Robert Maynard Hutchins ("Hutch" for short) for the recently departed chancellor who had been a hero to all the Chicago students. Chandra, growing impatient to see Bisshopp more often and not knowing what the problem was, abruptly asked him one day why he was so slow in moving to Williams Bay. The eager student told him about the reluctant landlords and how attached he was to Hutch. Chandra, taken aback, replied, "You have a dog, is it? Being attached to a wife I can understand, but sell the dog!" However, the story ended happily as Bisshopp was able to find a one-room apartment with a willing landlord soon after that little exchange, and keep his dog, although he probably had to pay a higher rent than he had planned.

When Chandra moved to the campus in 1964, the physics and astronomy students who were then working on their theses with him moved too. He had not been teaching at Yerkes for several years, and most of them, and the observational students of the time as well, had only taken a general relativity course from him, which he had taught in Chicago. They regarded it as more of a "cultural" course than one in which they would learn material that they might actually apply in research themselves. In fact, Chandra himself was getting seriously into the subject, and very soon began publishing in it. More than a decade earlier, around 1951, H. Lawrence Helfer had expressed interest in doing his Ph.D. thesis in general relativity, but Chandra said that he was not working in that field then, although he expected to do something in it when he neared retirement. Actually, of course, he did not wait that long, and he went on to do a great deal of research in general relativity, much of it after the "normal" retirement age. Donna Elbert had moved to the campus before him, in 1958, and continued working as Chandra's computer and secretary for many years there. She coauthored eighteen papers with him in all, and did the numerical computations for many more, as well as typing his papers and correspondence. She was a friend to all his graduate students and postdocs, but in 1979, as his human computing needs decreased, she moved on to become office manager for the astronomy department.

I have tried to compile a list of all of Chandra's Ph.D. thesis students through the last one he had, Bonnie D. Miller, who finished in 1973. Chandra had such a list, but I was unable to obtain a copy of it. Therefore I began with a tentative list of those I knew personally, and asked all of them, and also many other Ph.D.'s from Yerkes, to add whatever names they could to it. Proceeding iteratively in this way, I ended up with the list of forty-six given in appendix 1. Evidently I missed a few of his students, for more than one near the end of the list has written me that they were higher (by one or two) on Chandra's list. Partly it may be a matter of definition; some students had more than one thesis adviser, and I have included Anne Underhill (who had Chandra, Struve, and Jesse L. Greenstein as advisers) and Russell Kulsrud (Chandra, M. L. Goldberger, and Strömgren). There may have been others whom Chandra included on his list, but whom the rest of us did not perceive as his students. Of course, I would appreciate very much learning of any more of Chandra's Ph.D. thesis students whose names should be added.

It is striking to see in the table how closely the subjects of his students' theses tracked Chandra's own research. Whatever general subject he was working on (as described in other chapters in this memorial volume), his students were working on it too. He always had many problems which he could assign to students who were looking for thesis topics. These problems were hard but doable. He knew the subject well and could give good advice on how to proceed at each stage. Chandra never gave a student a problem he could not have done himself. Often they were more applied than he wanted to do, but they were all problems that, in some sense or other, needed doing. I never heard of a student of Chandra's who did not finish a Ph.D. thesis; he was a realist.

Furthermore, he was always interested in a student's thesis. He wanted to know what was going on, wanted to discuss the work, wanted to see progress. Although he was a great scientist, with many calls on his time, and the *Astrophysical Journal* was a constant drain of his energy for nineteen years, he was approachable to his students and would always make time for them somehow or other. He demanded a lot; nearly all of them commented that he made them work harder than they ever thought they would, but practically every one, looking back on it, thought it had been good for them. Chandra was an excellent, and highly productive, thesis adviser.

11.12   Chandra to his Ph.D. thesis students

I tried to get in touch with all of Chandra's thesis students to survey their thoughts on Chandra. Seven of them are no longer living: Gordon Wares, Ralph Williamson, Marjorie Harrison, Su-shu Huang, Marshal Wrubel, Frank Edmonds, and Nelson Limber. I sent all the others listed in appendix 1 a fairly long form letter; thirty-five of them, ninety per cent of those who are alive, responded. Overwhelmingly they thought that Chandra had been a good teacher for them. We may differ a little in just how mathematical he was, or how physical, and whether he would have been a little better if he had been a little more in one direction or the other, but we all believed he was very good indeed. One respondent even thought that any criticism of Chandra's teaching or research might simply be a sign of a deficiency in the criticizer, but all the rest considered him a human being!

Chandra liked students, but he would not tolerate any nonsense from them. Always formal, a bit reserved, well-dressed, he was also eager to discuss science, most approachable, and full of stories, especially of the great men of his youth and of *their* idiosyncrasies. In his later years Chandra became somewhat more aloof, and a larger fraction of the students who completed their degrees in the late 1960s and early 1970s considered him a bit cold or inhuman.

He was famous for inviting his students and postdocs to come over to his house for "some fun" on nice fall weekend days; this turned out to mean helping to rake leaves which had fallen from the huge trees which grew everywhere on the Yerkes grounds. With the wives of his students and colleagues he was unfailingly friendly and polite; to their children he was a kindly uncle-figure on the rare occasions when he saw them. To the men students he could be harsh if they did not measure up, but it was for their own good, he thought, and many of them agreed in retrospect. In his later years he evidently mellowed, for there are fewer reports along this line after he moved to Chicago. Perhaps this was an aspect of his aloofness. None of the women Ph.D.s who wrote me, his own students or the students of others, reported the slightest unkindness from Chandra; he was unfailingly friendly and polite with them. One, Anne Underhill, believed that he preferred women graduate students because "they tended to work hard all the time, while the men had sense enough to say, 'I have done enough; I will stop here.' "

In colloquia Chandra was quite capable of interrupting a speaker to criticize his work; he had dedicated his life to scientific truth and therefore felt it his duty to combat error. One former Chandra Ph.D. likened him to President Harry Truman, who stated, "I didn't give anybody hell, I just told the truth and they thought it was hell!" But this could be quite discomfiting to faculty colleagues whose work he criticized in each other's presence and the presence of their graduate students. Sometimes "the truth" was not so apparent to them as it was to Chandra.

He was tremendously supportive of all his Ph.D. students, recommending them for assistantships, fellowships, and, later, jobs, and following their progress with interest. Naturally, if he thought they were not working hard enough, or not choosing important enough problems, he did not hesitate to set them straight. In my own case I feel certain that he recommended me strongly for the predoctoral fellowship at Yerkes, the postdoctoral fellowship at Princeton, and my first faculty job at Caltech, which got me started in astrophysics. And in those bad old pre-open-recruitment, pre-search-committee days, a recommendation from Chandra meant a lot! Several years later, when I was in charge of the astronomy colloquia at Wisconsin, I managed to persuade him to drive up to Madison from Williams Bay and give one. I arranged for him to get a parking permit, and explained in detail just where he should pick it up and how important it was to display it on his car. When he arrived at our department office, I immediately asked him if he had found the campus police station and gotten the permit without any trouble. "No," he replied "I didn't bother. I just parked out in front!". "But you'll get a ticket!" I half-screamed. "If I do I'll just write Arthur D. Code [our chairman, another of his former students] on it," he calmly said; "What's the use of having friends if you don't use them?" Then I did not feel so bad about getting all those recommendations from him. A year or two later, soon after I became an associate professor, I wrote to ask Chandra some question about one of the methods in his book on radiative transfer. In his reply he joked that he had been surprised to hear from me, because most of his students stopped writing after they got tenure! As "Conversations with Chandra," the epilogue to Wali's book, shows, he was keenly aware of the academic hierarchy and rat race, and had a good, if slightly cutting, sense of humor.

Henry Horak, who taught at the University of Kansas for many years, had to persuade Chandra to travel much farther to give a colloquium there. He was understandably reluctant to make a special trip to do it, but finally, after many requests, told Horak he would come if he could gain an interview for Lalitha and himself with former President Harry Truman, then retired and living in Independence, Missouri. They had both become American citizens and liberal Democrats, and were active supporters of the party. No doubt Chandra thought that this condition would put a stop to Horak's importuning, but in fact a friend of his on the Kansas faculty was a ghost writer who had worked on Truman's memoirs. Thus he easily met the condition, so Chandra traveled west, gave the colloquium, and with Lalitha and Horak met Truman in the replica of the Oval Office at his presidential library. According to his former student, Chandra, awed, was for once almost speechless. So Horak, whose inclinations ran more toward the conservative side, did more of the talking than he had intended. Asked about his recall and relief of General MacArthur, Truman bluntly replied, "He disobeyed orders, and I fired him!" But then the former President went on to express a genuine appreciation of scientists, and they could all agree with that. Later, of course, Chandra accepted the National Medal of Science from President Lyndon B. Johnson in 1967.

Chandra was critical of a few of his former students, particularly ones who continued to do research in fields in which he had once worked, but had abandoned. I witnessed two such cases, in which my sympathies were all with his former students. They were not continuing along his lines, but were trying to go beyond him using their own, different methods. To my mind he was unnecessarily harsh with both of them, and did not show an open mind about examining their work on its merits. But he was just applying his own very high standards to them, he thought.

As the other chapters in this memorial volume describe, Chandra wrote numerous books, and, in particular, a series of major monographs summarizing his and his students' work on each of his successive fields of research as he left it for the next. One of his students, Surindar K. Trehan, compiled notes from a course in plasma physics which Chandra gave on campus in 1957–58 into a book. He approved highly of it, and the book was published by the University of Chicago Press (Trehan 1960). It eventually went through two reprintings, in 1962 and 1975. Chandra was very pleased with it, and especially liked to stress its analogy to the famous notes on Enrico Fermi's course on nuclear physics, compiled by three of his students and also published by the University of Chicago Press (Orear, Rosenfeld, and Schluter 1950). Trehan's book is a faithful rendition of Chandra's lecturing style, highly mathematical and somewhat uneven in its treatment of different topics, but containing some real gems of derivations. Quite naturally it is less polished than Chandra's magisterial summaries of a subject, but Trehan's book was immediately available to the many students and working scientists all over the world who were then hastening to learn as much about plasma physics as they could.

Perhaps the all-time story of Chandra's support of his students is that of Carl Rosenkilde. He was a physics graduate student on campus who admired Chandra's approach to theoretical subjects from taking his course on classical electrodynamics, in the years when he still had his office at Yerkes but was teaching on campus. Rosenkilde wanted to do a thesis with Chandra, and made an appointment to drive to Williams Bay to visit him and discuss a manuscript he had already written on the transmission of a charged particle through a kink in a magnetic field. But driving west from Kenosha on Wisconsin Highway 50, then a two-lane road, Rosenkilde's car was hit by another one and totaled. He suffered head injuries, and an ambulance took him to a medical clinic near the crash scene to be examined and treated. Dazed and half-conscious, Rosenkilde was worried about keeping Chandra waiting, and conveyed the information to someone that he was on his way to visit him. While he was still being examined, Rosenkilde heard Chandra's distinctive voice outside in the waiting room, asking if he were still alive! He had gotten the news by phone, and came as soon as he could to help out. Relieved to find Rosenkilde not only alive but conscious and rapidly improving, Chandra insisted on driving him back to the observatory, telling him on the way a story from India about the Grim Reaper's early arrival on some other occasion. When he got to his office with Rosenkilde, Chandra would not allow him to discuss his manuscript, but instead drove him to the station and put him on the next train back to Chicago, accompanied by another of his physics graduate students, Lawrence Lee. However, Chandra kept the manuscript, gave it to a referee, and, receiving a favorable report on it, published it in the *Astrophysical Journal* (Rosenkilde 1965). It was his first publication, and it convinced Chandra to accept him as a thesis student. Luckily for Rosenkilde, Chandra moved to Chicago soon after that, and the young graduate student never had to drive back to Yerkes again!

Chandra called upon many of his students to check his papers or books before publication. In my own case, for my last two years I was his senior graduate student at Yerkes after Horak finished his degree and Guido Münch, his thesis student earlier, and then his colleague on the faculty, left for a position at Caltech. When Chandra had finished writing a paper, Donna Elbert would type it for him, simultaneously making several carbon copies. He would fill in the complete equations on the original in ink, as well as the mathematical symbols in the text, all in his bold, characteristic writing. Then he gave me the original and the carbons; I filled in the equations and symbols on them at the same time I checked through the mathematical manipulations step by step. Presumably there would only be misprints in signs or omissions of a symbol by that point, but he cautioned me to check each step carefully. For two years I did, but I almost never found an error; I think it proved that he did not make them, but I could not be sure. Many of his other thesis students had similar assignments, and those who were working with him when he wrote a book checked the equations in his manuscript, and later the proofs. It was good practice for the future.

All of Chandra's Ph.D. thesis students enjoyed working with him. They learned to do research under a master of it, and those who stayed in his general area of highly mathematical theoretical astrophysics were masters of it too by the time they had completed their theses. He taught, seasoned, encouraged, and broadened all of us while we were his students, and he supported us greatly after we had left the nest.

11.13 Postdocs and parody

Besides the many Ph.D. thesis students he trained, Chandra had quite a few postdoctoral research associates who collaborated with him over the years. I have not made a systematic attempt to obtain their stories, but can briefly mention a few of them. One was Mario Schoenberg, who came to America from the University of São Paulo, Brazil with a Guggenheim Fellowship. Under it he worked at George Washington University with George Gamow on neutrino cooling of dense, hot stellar objects by what they named the Urca process, which they proposed as the mechanism for initiating supernovae, an idea that has lasted very well. Then Schoenberg moved on to Yerkes, and with Chandra worked on the evolution of stellar models with burnt-out, gaseous, isothermal cores (Schoenberg and Chandrasekhar 1942). Their paper followed one published the previous year by Louis Henrich and Chandra (not part of Henrich's thesis). It analyzed mathematically the idea of Gamow and Edward Teller that red giants are stars in which D-, Li-, Be-, and B-burning in shells just outside an isothermal core in which these light nuclei have already been exhausted are the main energy sources (Henrich and Chandrasekhar 1941). They discovered the upper limit to the fraction of the mass which could exist in a burnt-out nuclear core, now called the "Schoenberg-Chandrasekhar limit." That later paper in fact extended this concept to include a discontinuity in molecular weight, and carefully traced the evolution of shell-source stars burning H rather than light elements. It was one of the key first steps toward the recognition of the true nature of red-giant stars as late stages of normal stellar evolution.

With these two papers and Gordon Wares's thesis on partially degenerate models, completed in 1940, Chandra was very much in the thick of the beginning of the study of stellar evolution (Wares 1944). In 1938 and 1939 he had taken part in the very important conferences on nuclear energy production, organized in Washington by Gamow and Merle A. Tuve. Then in 1941, with Henrich as his assistant, Chandra made a pioneering study of the equilibrium distribution of nuclear abundances at very high temperatures and densities, related to what we now call the r-process in supernovae. In it they supposed that "the" abundances of the elements were fixed under prestellar conditions in an expanding universe, and by fitting the abundance and isotope ratios then considered universal, derived $T \approx 8 \times 10^9 \, ^\circ\mathrm{K}$ and $\rho \approx 10^7 \, \mathrm{gm/cc}$ as the conditions under which the elements O through Si had formed (Chandrasekhar and Henrich 1942). They recognized that the iron-peak elements could not be formed under these same conditions, but stated that perhaps they had frozen out under earlier, even more extreme conditions. There are many similarities between their "scenario" (in words of today) and our current ideas of the formation of the α-element nuclei in supernovae, including their estimate of the mean temperature and density at which these elements were made.

But, as the study of stellar interiors and evolution was becoming more physical and detailed, Chandra was shifting his research to more mathematical stochastic and statistical problems. Ironically enough, he did some of this latter work jointly with John von Neumann, the great exponent and facilitator of applications of nuclear fission and fusion chain reactions in the "real world." Chandra himself next went on to radiative transfer, and left further developments in understanding stellar evolution to Martin Schwarzschild, Fred Hoyle, Louis G. Henyey, and their co-workers.

In the late 1950s Chandra had an especially large and active group of postdocs working with him at Yerkes on plasma physics and hydromagnetic stability. One was Lo Woltjer, who had completed his brilliant thesis on the Crab Nebula at Leiden Observatory with Jan H. Oort, earning his Ph.D. in 1957 and then coming to America to work with Chandra. Others were John Hazlehurst, Paul H. Roberts (who later joined the Yerkes faculty for a time) and John Sykes from England, René Simon from Belgium, and William H. Reid, an American who had done his Ph.D. at Trinity College, Cambridge with Ian Proudman. Kevin Prendergast and Nelson Limber, then young Yerkes faculty members, were working with them on some of these problems. In addition Dave Fultz, a geophysicist, and Russell Donnelly, a physicist, both faculty members on the campus, were involved with him through their experimental work on hydrodynamic and hydromagnetic stability, along with Yoshio Nakagawa, a research associate.

From this period dates the famous parody "On the imperturbability of elevator operators. LVII," by "S. Candlestickmaker," printed as a reprint from the *Astrophysical Journal.* Full of outrageous puns, double entendres, overstatements, and inside jokes, it is written in a style reminiscent of one of Chandra's papers, but with all his idiosyncrasies greatly exaggerated. In fact the Candlestickmaker parody is closely modeled on a paper by Chandra entitled "The instability of a layer of fluid heated below and subject to the simultaneous action of a magnetic field and rotation. II." It had appeared in the *Proceedings of the Royal Society of London* (of which Chandra was a Fellow) the previous year, and the twenty references in the parody to previous papers by the purported author, from "S. Candlestickmaker (1954a)" up to "(1954t)," are attributed to different actually existing but somewhat implausibly named journals such as the *Transactions of the North-east Coast Institute of Engineers and Shipbuilders,* **237**, 476, *The Journal of Dairy Science,* **237**, 476, and *Scientific Progress of the Twentieth Century,* **237**, 476, all with the same volume and page numbers as the real paper (Chandrasekhar 1956). There is also one reference to a paper by Candlestickmaker and Miss Canna E. Helpit (Donna D. Elbert), whom the author thanks for her laborious numerical work in obtaining the approximate solution for the single case in which the problem can be solved explicitly, "admittedly a case which has never occurred in living memory," but "from past experience with problems of this kind one may feel that any solution is better than none." The equation for this case is

$$\ln \Omega_{2l} = 1,$$

and the approximate solution is given as

$$\Omega_{2l} = 2.7.$$

The date the paper was received is given as October 19, 1910, Chandra's date of birth, and the institution of the author is stated to be the "Institute for Studied Advances, Old Cardigan, Wales."

The paper has to be read carefully to be fully appreciated; it has been republished in the *Quarterly Journal of the Royal Astronomical Society* (Sykes 1972) and in an anthology edited by Weber (1973). However, these versions as printed do not do justice to the original, which was widely circulated as a "reprint" from the *Astrophysical Journal,* printed exactly in its style, typeface and format, as if it had been published in volume **237**, number 1211, November 1957. Of course the first page is numbered 476.

The month and year are when the "reprint" was actually printed, but the volume number then seemed so large that it would appear impossibly far in the future (**126** was then current). In fact, that volume number appeared in 1980, and when Chandra died in August 1995 the *Astrophysical Journal* was publishing volume **450**.

The author of the parody was John Sykes, listed as the member who had "communicated" the paper, although no doubt some of the other postdocs contributed additional touches to it. Sykes, a brilliant mathematician, linguist, translator, and puzzle solver, was working with Chandra as a postdoc for a year to get into magnetohydrodynamics, before returning to Harwell to join the United Kingdom fusion project. Soon afterward, however, he became a full-time translator and lexicographer. Sadly, he died of a heart attack a few years ago. In 1957 Sykes submitted the parody by mail to the *Astrophysical Journal* office as an ordinary manuscript intended for publication. The secretary who opened it recognized or at least suspected that it was a joke and took it to Chandra, fearing that he might explode. But he was delighted with it, and showed it to everyone who came into his office. He authorized printing it in the *Astrophysical Journal* reprint format, and the postdocs and students, headed by William Reid, George Backus, and Kumar Trehan, took up a collection to pay for it. Everyone who saw the reprint was amused by it; the more closely they had studied his papers, the better they understood some of the allusions in it. Chandra thought it was a wonderful joke, but he also recommended it seriously to more than one of his students as a good example of the correct style in which to write a scientific paper. Evidently Sykes had captured his style nearly perfectly.

11.14 Conclusion

Chandra was a great scientist, who was also an excellent teacher and thesis adviser of graduate students. He was an extremely productive research worker, who published a prodigious number of scientific papers and research monographs. In addition, he guided a very large number of graduate students to their Ph.D. degrees and started them on their own research careers. He supported most of them with praise and recommendations in their later lives. To many scientists outside Yerkes Observatory and the University of Chicago, Chandra seemed a remote, forbidding figure. But to his own graduate students he was highly approachable, even outgoing. All the graduate students who worked with him at Yerkes Observatory and on the University of Chicago campus up through 1973 felt that they had learned much from him, and had been fortunate to have been his students. A few thought of him as a god; most recognized him as an exceptional human being.

I am very grateful to the many former graduate students, most of them ones who received their Ph.D.s at Yerkes and on the campus, for their letters, e-mail messages, and phone calls in response to my request for information on their memories of Chandra and their interactions with him. I am especially grateful to Peter O. Vandervoort, Bonnie D. Miller, Henry G. Horak, Donat G. Wentzel, Philip J. Greenberg, and John L. Friedman, who all played major roles in filling in the list of names of Chandra's Yerkes Ph.D. students given in appendix 1. I am indebted to Maxine Hunsinger Sullivan, University of Chicago Registrar, for providing the names of all the students who were registered for Chandra's astrophysics course on the campus in 1948, even though, as she wrote me, she would have preferred to preserve the legend. I also wish to thank Chen Ning Yang, Tsung-Dao Lee, Richard L. Garwin, and Arthur Uhlir for their individual recollections of this class we all attended. I am most grateful to Donna D. Elbert for her memories of some of Chandra's interactions with his students, to William H. Reid for his account of some of the particulars of the writing and "reprinting" of the S. Candlestickmaker parody, and to Roger Tayler for his recollections of what the late John Sykes had told him about his role in it.

References

Abt, H. A. 1995a, *Astrophys. J.*, **454**, 551; 1995b, *Astrophys. J.*, **455**, 407.

Aizenman, M. 1968, *Astrophys. J.*, **153**, 511.

Backus, G. E. 1957, *Astrophys. J.*, **125**, 500.

Bisshopp, F. E. 1958, *Phil. Mag.*, **3**, 1342; 1962, *J. Math. Mech.*, **11**, 667.

Bonanos, S. 1971, *Commun. Math. Phys.*, **22**, 190.

Buti, B. 1962, *Phys. Fluids*, **5**, 1; 1963a, *Phys. Fluids*, **6**, 89; 1963b, *Phys. Fluids*, **6**, 100.

Chandrasekhar, S. 1947, *Science*, **106**, 213; 1949, *Astrophys. J.*, **110**, 329; 1956, *Proc. R. Soc. London*, **A237**, 476; 1964, *Proc. Am. Phil. Soc.*, **108**, 1.

Chandrasekhar, S., Fermi, E. 1953a, *Astrophys. J.*, **118**, 113; 1953b, *Astrophys. J.*, **118**, 116.

Chandrasekhar, S., Friedman, J. L. 1971, *Phys. Rev. Lett.*, **261**, 1047; 1972a, *Astrophys. J.*, **175**, 379; 1972b, *Astrophys. J.*, **176**, 745; 1972c, *Astrophys. J.*, **177**, 745; 1973a, *Astrophys. J.*, **181**, 481; 1973b, *Astrophys. J.*, **185**, 1.

Chandrasekhar, S., Henrich, L. R. 1942, *Astrophys. J.*, **95**, 288.

Chandrasekhar, S., Krogdahl, M. K. 1943, *Astrophys. J.*, **98**, 205.

Chandrasekhar, S., Nutku, Y. 1969, *Astrophys. J.*, **158**, 55.

Chandrasekhar, S., Wares, G. W. 1949, *Astrophys. J.*, **109**, 551.

Chandrasekhar, S., Wright, J. P. 1961, *Proc. Nat. Acad. Sci.*, **47**, 341.

Clement, M. J. 1964, *Astrophys. J.*, **140**, 1045; 1965a, *Astrophys. J.*, **141**, 210; 1965b, *Astrophys. J.*, **141**, 1443; 1965c, *Astrophys. J.*, **142**, 243.

Code, A. D. 1950, *Astrophys. J.*, **112**, 22.

Conwell, E. M. 1948a, *Phys. Rev.*, **74**, 268; 1948b, *Phys. Rev.*, **74**, 277.

Edmonds, F. N. 1950a, *Astrophys. J.*, **112**, 307; 1950b, *Astrophys. J.*, **112**, 324.

Esposito, F. P. 1971, *Astrophys. J.*, **165**, 165.

Friedman, J. L. 1973, *Proc. R. Soc. London*, **A335**, 163.

Greenberg, P. J. 1971a, *Astrophys. J.*, **164**, 569; 1971b, *Astrophys. J.*, **164**, 589.

Harrison, M. H. 1946a, *Astrophys. J.*, **102**, 216; 1946b, *Astrophys. J.*, **103**, 193; 1947, *Astrophys. J.*, **105**, 322.

Helfer, H. L. 1953, *Astrophys. J.*, **117**, 177; 1954, *Astrophys. J.*, **119**, 34.

Henrich, L. R. 1941, *Astrophys. J.*, **93**, 483; 1942, *Astrophys. J.*, **96**, 106; 1943, *Astrophys. J.*, **98**, 192.

Henrich, L. R., Chandrasekhar, S. 1941, *Astrophys. J.*, **94**, 525.

Hiltner, W. A., 1953, Chandrasekhar, S. 1945, *Astrophys. J.*, **102**, 135.

Horak, H. G. 1950, *Astrophys. J.*, **112**, 445.

Huang, S. S. 1948, *Astrophys. J.*, **108**, 354; 1950, *Astrophys. J.*, **112**, 418.

Jensen, E. 1955, *Astrophys. J. Suppl.*, **2**, 141.

Krefetz, E. 1967a, *Astrophys. J.*, **148**, 589; 1967b, *Astrophys. J.*, **148**, 613.

Kristian, J. 1963a, *Astrophys. J.*, **137**, 102; 1963b, *Astrophys. J.*, **137**, 117.

Krogdahl, W. 1942, *Astrophys. J.*, **96**, 124.

Krogdahl, M. K. 1944a, *Astrophys. J.*, **100**, 311; 1944b, *Astrophys. J.*, **100**, 333.

Kulsrud, R. M. 1955, *Astrophys. J.*, **121**, 461.

Lebovitz, N. R. 1960, *Proc. Cambridge. Phil. Soc.*, **56**, 154; 1961a, *Proc. Cambridge. Phil. Soc.*, **57**, 583; 1961b, *Astrophys. J.*, **134**, 500.

Lee, E. P. 1968, *Astrophys. J.*, **151**, 687.

Lee, L. L. 1965, *Annals Phys.*, **32**, 292.

Limber, D. N. 1953a, *Astrophys. J.*, **117**, 134; 1953b, *Astrophys. J.*, **117**, 145; 1954, *Astrophys. J.*, **119**, 655.

Miller, B. D. 1973, *Astrophys. J.*, **181**, 497; 1974, *Astrophys. J.*, **187**, 609.

Münch, G. 1945, *Astrophys. J.*, **102**, 385; 1946, *Astrophys. J.*, **104**, 87.

Nduka, A. 1971, *Astrophys. J.*, **170**, 131.

Nutku, Y. 1969a, *Astrophys. J.*, **155**, 999; 1969b, *Astrophys. J.*, **158**, 991.

Orear, J., Rosenfeld, A. H., Schluter, R. A. 1950, *Nuclear Physics: A Course Given by Enrico Fermi at the University of Chicago* (Chicago: University of Chicago Press)

Osterbrock, D. E. 1951, *Astrophys. J.*, **114**, 469; 1952a, *Astrophys. J.*, **116**, 164; 1952b, *Phys. Rev.*, **87**, 468; 1995, *Astrophys. J.*, **438**, 1; 1997, *Yerkes Observatory 1892-1950: The Birth, Near Death, and Resurrection of A Scientific Research Institution* (Chicago: University of Chicago Press).

Ostriker, J. 1964a, *Astrophys. J.*, **137**, 1056; 1964b, *Astrophys. J.*, **137**, 1067; 1964c, Ãstrophys. J., **137**, 1529; 1965, *Astrophys. J. Suppl.*, **11**, 167.

Persides, S. 1970, *Proc. R. Soc. London*, **A320**, 349.

Rosenkilde, C. E. 1965, *Astrophys. J.*, **141**, 1105; 1967a, *J. Math. Phys.*, **8**, 84; 1967b, *J. Math. Phys.*, **8**, 88; 1967c, *J. Math. Phys.*, **8**, 98.

Rossner, L. F. 1967, *Astrophys. J.*, **149**, 145.

Schoenberg, M., Chandrasekhar, S. 1942, *Astrophys. J.*, **96**, 161.

Stettner, R. 1971a, *Annals Phys.*, **67**, 238; 1971b, *Annals Phys.*, **68**, 281; 1971c, *J. Math. Phys.*, **12**, 1159.

Sykes, J. 1972, *Q. J. R. Astr. Soc.*, **13**, 63.

Trehan, S. K. 1957, *Astrophys. J.*, **126**, 429; 1958, *Astrophys. J.*, **127**, 436; 1960, *Plasma Physics: A Course Given by S. Chandrasekhar at the University of Chicago* (Chicago: University of Chicago Press).

Tuberg, M. 1946, *Astrophys. J.*, **103**, 145.

Underhill, A. B. 1947, *Astrophys. J.*, **106**, 128; 1948a, *Astrophys. J.*, **107**, 247; 1948b, *Astrophys. J.*, **107**, 349.

Vandervoort, P. O. 1960, *Annals Phys.*, **10**, 401.

Wali, K. S. 1991, *Chandra, A Biography of S. Chandrasekhar* (Chicago: University of Chicago Press).

Wares, G. W. 1944, *Astrophys. J.*, **100**, 158.

Weber, R. L. ed., 1973, *A Random Walk in Science* (London and Bristol: The Institute of Physics), 100.

Wentzel, D. G. 1960, *Astrophys. J. Suppl.*, **5**, 137.

Williamson, R. E. 1942, *Astrophys. J.*, **96**, 438; 1943, *Astrophys. J.*, **97**, 51.

Wright, J. P. 1960, *Phys. Fluids*, **3**, 607; 1961, *Phys. Fluids*, **4**, 1341.

Wrubel, M. H. 1949a, *Astrophys. J.*, **109**, 66; 1949b, *Astrophys. J.*, **110**, 288; 1950a, *Astrophys. J.*, **111**, 157; 1950b, *Astrophys. J.*, **112**, 424.

Appendix 1

Chandra's Ph.D. Students at Yerkes Observatory
and in Chicago: Thesis Topics and References.

Name	Ph.D. degree	Thesis topic / Reference(s)
Gordon W. Wares	1940	Partially degenerate stellar models (Wares 1944; Chandrasekhar and Wares 1949)
Louis R. Henrich	1942	Radiation pressure in stellar models (Henrich 1941, 1942, 1943)
Wasley S. Krogdahl	1942	Rotational distortion of stellar models (Krogdahl 1942)
Ralph E. Williamson	1943	H^- absorption coefficient and continuous spectra of model stellar atmospheres (Williamson 1942, 1943)
Margaret Kiess Krogdahl	1944	Interaction of H^+ with H and He and resulting Stark effect in stellar atmospheres (Krogdahl 1944a,b)
Guido Münch	1946	Continuous spectrum of the Sun and agreement of derived absorption coefficient with H^- (Münch 1945, 1946)
Merle E. Tuberg	1946	Variation of solar absorption-line contours across the disk of the sun (Tuberg 1946)
Marjorie Hall Harrison	1947	Composite stellar models with partly-degenerate isothermal or gravitationally contracting cores (Harrison 1946a,b; 1947)

Name	Ph.D. degree	Thesis topic / Reference(s)
Anne B. Underhill	1948	Theory and observation of absorption-line profiles in B stars (Underhill 1947; 1948a,b)
Esther Conwell	1948*	Improved wave functions for O^-, H^- (Conwell 1948a,b)
Su-shu Huang	1949	He absorption coefficient (Huang 1948)
Marshal H. Wrubel	1949	Curves of growth; radiative transfer in a spherical, electron-scattering atmosphere (Wrubel 1949a,b; 1950a)
Arthur D. Code	1950	Radiative equilibrium in an atmosphere with combined scattering and pure absorption (Code 1950)
Frank N. Edmonds Jr.	1950	Radiative transfer in an expanding, spherical, electron-scattering atmosphere (Edmonds 1950a,b)
Henry G. Horak	1950	Diffuse reflection from planetary atmospheres (Horak 1950)
Donald E. Osterbrock	1952	Time of relaxation of a star in a fluctuating density field; transition probabilities of forbidden lines; electrical conductivity (Osterbrock 1951; 1952a,b)
H. Lawrence Helfer	1953	Magnetohydrodynamical shock waves and finite-amplitude waves in an infinite homogeneous medium (Helfer 1953, 1954)
D. Nelson Limber	1953	Analysis of galaxy counts in terms of absorption by a fluctuating density field (Limber 1953a,b; 1954)

Name	Ph.D. degree	Thesis topic / Reference(s)
Eberhart Jensen	1953	Magnetohydrodynamic oscillations of a conducting, fluid sphere in a magnetic field (Jensen 1955)
Russell M. Kulsrud	1954*	Effect of a magnetic field on turbulent generation of noise (Kulsrud 1955)
George E. Backus	1956*	Nonexistence of axisymmetric fluid dynamos (Backus 1957)
Surindar K. Trehan	1958*	Stability of force-free magnetic fields (Trehan 1957, 1958)
Fred E. Bisshopp	1959*	Thermal convection in a rotating fluid sphere (Bisshopp 1958, 1962)
Peter O. Vandervoort	1960*	Relativistic motion of a charged particle in an inhomogeneous electromagnetic field (Vandervoort 1960)
Donat G. Wentzel	1960*	Hydromagnetic equilibria (Wentzel 1960)
Norman R. Lebovitz	1961*	Equilibrium stability of a system of disk dynamos; virial tensor and its application to self-gravitating fluids (Lebovitz 1960; 1961a,b)
James P. Wright	1961**	Effect of neutral particles on transport properties of a plasma in a magnetic field; general relativity (Wright 1960, 1961; Chandrasekhar and Wright 1961)
Jerome Kristian	1962*	Hydromagnetic oscillations about equipartition (Kristian 1963a,b)

Name	Ph.D. degree	Thesis topic / Reference(s)
Bimla Buti	1962*	Relativistic effects on plasma oscillations and two-stream instabilities (Buti 1962; 1963a,b)
Jeremiah P. Ostriker	1964	Equilibrium, stability and oscillations of self-gravitating cylindrical configurations (Ostriker 1964a,b,c; 1965)
Lawrence L. Lee	1964*	Magnetohydrodynamic turbulence in an incompressible fluid (Lee 1965)
Maurice J. Clement	1965	Nonradial oscillations of rotating stars (Clement 1964; 1965a,b,c)
Carl E. Rosenkilde	1966*	Stability of axisymmetric rotating charged liquid-drop nuclear models (Rosenkilde 1967a,b,c)
Lawrence F. Rossner	1966	Finite-amplitude oscillations of Maclaurin spheroids (Rossner 1967)
Elliott I. Krefetz	1966*	Slowly rotating ideal-fluid configurations in the post-Newtonian approximation (Krefetz 1967a,b)
Morris L. Aizenman	1967	Roche-Riemann ellipsoids (Aizenman 1968)
Edward P. Lee	1968*	Brownian motion in a stellar system (Lee 1968)
Yavuz Nutku	1969*	Post-Newtonian hydrodynamics equations and energy-momentum complex in general relativity and Brans-Dicke theory (Nutku 1969a,b; Chandrasekhar and Nutku 1969)

Name	Ph.D. degree	Thesis topic / Reference(s)
Philip J. Greenberg	1970*	Post-Newtonian equations of hydrodynamics and of magnetohydrodynamics in general relativity (Greenberg 1971a,b)
Sotirios C. Persides	1970*	Asymptotically flat gravitational fields in general relativity (Persides 1970)
F. Paul Esposito	1971*	Absorption of gravitational energy by a viscous compressible fluid (Esposito 1971)
Sotirios Bonanos	1971*	Stability of the Taub universe (Bonanos 1971)
Roger Stettner	1971*	Post-Newtonian approximation for electromagnetism (Stettner 1971a,b,c)
Amagh Nduka	1972*	Roche problem in an eccentric orbit (Nduka 1971)
John L. Friedman	1973*	Stability of rotating stars in general relativity (Chandrasekhar and Friedman 1971; 1972a,b,c; 1973a,b; Friedman 1973)
Bonnie D. Miller	1973	Effects of gravitational radiation on the evolution and secular stability of stars (Miller 1973, 1974)

* Indicates Ph.D. in physics.
** Indicates Ph.D. in chemistry.
No symbol indicates Ph.D. in astronomy and astrophysics.

Contributors

James Binney
  Theoretical Physics
  Oxford University
  Keble Road, Oxford OX1 3NP, UK

John L. Friedman
  Department of Physics
  University of Wisconsin
  Milwaukee, WI 53201, USA

Norman R. Lebovitz
  The University of Chicago
  Chicago, IL 60637, USA

Donald E. Osterbrock
  Lick Observatory
  University of California
  Santa Cruz, CA 95064, USA

E. N. Parker
  Enrico Fermi Institute and Department of Physics and Astronomy
  University of Chicago
  Chicago, IL 60637, USA

Roger Penrose
  Mathematical Institute
  University of Oxford
  Oxford OX1 3LB, UK

A. R. P. Rau
  Department of Physics and Astronomy
  Louisiana State University
  Baton Rouge, LA 70803-4001, USA

George Rybicki
  Harvard-Smithsonian Center for Astrophysics
  60 Garden Street
  Cambridge, MA 02138, USA

E. E. Salpeter
  Cornell University
  Ithaca, NY 14853, USA

Bernard F. Schutz
  Max Planck Institute for Gravitational Physics
  The Albert Einstein Institute
  Potsdam, Germany

G. Srinivasan
  Raman Research Institute
  C. V. Raman Avenue
  Bangalore 560 080, India